Mathematical Aspects of Superspace

NATO ASI Series

Advanced Science Institutes Series

A series presenting the results of activities sponsored by the NATO Science Committee, which aims at the dissemination of advanced scientific and technological knowledge, with a view to strengthening links between scientific communities.

The series is published by an international board of publishers in conjunction with the NATO Scientific Affairs Division

A	Life Sciences	Plenum Publishing Corporation
B	Physics	London and New York
C	Mathematical and Physical Sciences	D. Reidel Publishing Company Dordrecht, Boston and Lancaster
D	Behavioural and Social Sciences	Martinus Nijhoff Publishers
E	Engineering and Materials Sciences	The Hague, Boston and Lancaster
F	Computer and Systems Sciences	Springer-Verlag
G	Ecological Sciences	Berlin, Heidelberg, New York and Tokyo

Series C: Mathematical and Physical Sciences Vol. 132

Mathematical Aspects of Superspace

edited by

H.-J. Seifert

Hochschule der Bundeswehr Hamburg, Hamburg, F.R.G.

C. J. S. Clarke

University of York, York, U.K.

and

A. Rosenblum

Temple University, Philadelphia, Pennsylvania, U.S.A.

D. Reidel Publishing Company

Dordrecht / Boston / Lancaster

Published in cooperation with NATO Scientific Affairs Division

Proceedings of the NATO Advanced Research Workshop on
Mathematical Aspects of Superspace
Hamburg, F.R.G.
12-16 July, 1983

Library of Congress Cataloging in Publication Data

NATO Advanced Research Workshop on Mathematical Aspects of Superspace (1983 : Hamburg, Germany)
Mathematical aspects of superspace.

(NATO ASI series. Series C, Mathematical and physical sciences ; v. 132)
"Proceedings of the NATO Advanced Research Workshop on Mathematical Aspects of Superspace, Hamburg, F.R.G., July 12–16, 1983"–CIP t.p. verso.
Bibliography: p.
Includes index.
1. Supersymmetry–Congresses. 2. Supergravity–Congresses. 3. Gauge fields (Physics)–Congresses. I. Seifert, Hans-Jürgen, 1942- II. Clarke, C. J. S., 1946- III. Rosenblum, A., 1943- IV. Title. V. Title: Superspace. VI. Series.
QC174.17.S9N38 1983 530.1'4 84-13353
ISBN-13: 978-94-009-6448-8 e-ISBN-13: 978-94-009-6446-4
DOI:10.1007/978-94-009-6446-4

Published by D. Reidel Publishing Company
P.O. Box 17, 3300 AA Dordrecht, Holland

Sold and distributed in the U.S.A. and Canada
by Kluwer Academic Publishers,
190 Old Derby Street, Hingham, MA 02043, U.S.A.

In all other countries, sold and distributed
by Kluwer Academic Publishers Group,
P.O. Box 322, 3300 AH Dordrecht, Holland

D. Reidel Publishing Company is a member of the Kluwer Academic Publishers Group

TABLE OF CONTENTS

PREFACE

Over the past five years, through a continually increasing wave of activity in the physics community, supergravity has come to be regarded as one of the most promising ways of unifying gravity with other particle interaction as a finite gauge theory to explain the spectrum of elementary particles. Concurrently important mathematical works on the arena of supergravity has taken place, starting with Kostant's theory of graded manifolds and continuing with Batchelor's work linking this with the superspace formalism.

There remains, however, a gap between the mathematical and physical approaches expressed by such unanswered questions as, does there exist a superspace having all the properties that physicists require of it? Does it make sense to perform path-integral in such a space? It is hoped that these proceedings will begin a dialogue between mathematicians and physicists on such questions as the plan of renormalisation in supergravity.

The contributors to the proceedings consist both of mathematicians and relativists who bring their experience in differential geometry, classical gravitation and algebra and also quantum field theorists specialized in supersymmetry and supergravity.

One of the most important problems associated with supersymmetry is its relationship to the elementary particle spectrum. The first question is why, at present, do we not experimentally see all the supersymmetric partners. One challenge in realistic model building is to make the right members heavy and experimentally unobservable. A mathematical framework for building theories which could treat this problem is provided by the contribution of Wess. He first introduces the supersymmetry algebra and the extended supersymmetry algebra. The transformation law of the Akulov-Volkov field is given as the transformation of a surface in super space. Finally, the superfield associated with the Akulov-Volkov Field is derived. The Akulov-Voltov field represents a non-linear realization of supersymmetry. Though not presented at the conference, the non-linear realizations of supersymmetry are very useful in model building as shown in the work of Wess and S. Samuel.

The mathematical framework underlying both conventional gravity theory and the superspace formulation in the context of modern differential geometry is presented by Clarke. He begins with the concept of a manifold, He then goes on to the concept of a fibre bundle and explains how gauge theories can be constructed using this idea and that of a connection term. Gauge invariance is then formulated in a differential geometric language. Spinors are then treated and linked with supersymmetric transformation. Clarke's

H.-J. Seifert et al. (eds.), Mathematical Aspects of Superspace, ix–xii.

article will, we hope, act as a guide for physicists for the more mathematical articles that follow in the proceedings.

A quantum field theory is one where the integrals of the field and its derivatives are already operators on a complex Hilbert space of quantum-mechanical "states", i.e. operator-valued distribution. In all of these lectures this is not the case, so we are dealing with fields that are non-quantum, even though not classical in the sense that their values at each point can be determined by an actual physical measurement. For Fermion field, we are dealing with anti-commuting c-numbers or Grassmann variables. Since all the envisaged applications of supergravity are in the quantum realm, one needs a method of quantising the non-quantum field. The most popular method is that of path integrals (Abers & Lee for gauge theories generally; Nicolai for supergravity). But the rigorous theory of path integrals has a long way to go. The theory explained by Elworthy in these proceedings forms a bridgehead for the eventual spanning of the gulf that still exists between the non-quantum theory and a future quantum theory. He presents a mathematical description of standard material on path integration on $\mathbb{R}^n$ and on finite dimensional Riemannian manifolds. Both Wiener and Feynman path integrals are discussed and their similarity is brought out both for paths on $\mathbb{R}^n$ and on manifolds M. A major aim of the presentation is to give a general setting which is likely to be of use in a variety of situations, in particular for path integration on superspace and supermanifolds. From a rigorous point of view, physical interpretations of supergravity are therefore restricted to inferences about the (as yet non-existent) quantum theory that is based on the non-quantum theory from which it must be derived, and which do not depend on the details of how one builds the quantum theory.

Because of uncertainty as to how the quantum theory is to be constructed, there is uncertainty about which non-quantum theory should be the starting point. By definition of "supergravity" the theory must be based on one of the super-Lie groups that extend the Poincaré group - probably by admitting a super-analogue of the gauge group in ordinary gauge theory. But here agreement on the most promising form of theory stops. Of course, a major consideration is the experimental agreement with the elementary particle spectrum.

One approach, that of Isenberg and his collaborators, tries to stay as close to the formalism of a traditional gauge-theory as possible. In particular he treats the supergravity version of the Gauchy initial value problem. This approach has the advantage that one draws on the experience gained in ordinary gauge theory to guide one as to which a non-quantum theory will produce a physically acceptable theory when quantized. But the minimal requirements of theories of this type (the anti-commutativity of the fields and the relation to the Lorentz groups) do not make explicit what the mathematical structure of the fields is: in which space are the functions describing the fields living. The simplest example would involve working in the bundle Ext(V), the exterior product bundle based on a vector bundle V, for example the bundle of Rarita-Schwinger fields. But is this the only possibility? For example, an approach of Tucker & Benn appears to use a system that is algebraically different as well as having a different representation. Unfortunately all this is also connected with the practical problems associated with path integrals for Fermions.

Two arguments point to a different approach, using graded manifolds. One is the need just mentioned for understanding what is the range of acceptable mathematical models availabe. The other is the argument that a representation of the fields as section of Ext(V) implies that one can distinguish globally which fields are purely bosonic and which are purely fermionic, a situation analogous to working in a particular gauge in a gauge theory. To express the full freedom of local supersymmetry transformations, one needs to use a system where the representations of the fields as an exterior algebra can be different in different local neighbourhoods, the representations in overlapping neighbourhoods being related by a supersymmetry transformation. This is precisely Kostant's sheaf-theoretic graded manifold. The content of Batchelor's theorem (see her lecture and also Rawnsley's contribution) is that, despite the difference in formalism between a graded manifold and an exterior bundle, one does not actually lose anything; every graded manifold can be represented in terms of sections of an exterior bundle.

The advantage of the graded manifold approach, which uses sheaf theory, is that it expresses the intrinsic algebraic structure of the fields as purely as possible, without assuming any particular representation. This simplifies the proof of general results concerning symmetries, covariance etc., and also opens the possibility of choosing different concrete representations for different purposes. More speculatively, it could assist with quantisation, since there are many approaches to quantum field theory where the basic physical objects are local operator algebras (von Neumann algebras associated with each open set) which have a sheaf-theoretic structure close to that used in the graded manifold approach explained by Batchelor.

It is characteristic of supersymmetries in the exterior bundle approach that they are not just "super" versions of gauge transformations, because a gauge transformation keeps the point of the manifold fixed, altering the field at each point, whereas a supersymmetry transformation moves the points of the manifold. This is usually expressed in terms of the "infinitesimal symmetry transformations" (derivations of the fields) where there appear Lie derivations of the field with respect to vector fields (on the bundle of values for the field) having a horizontal component, not tangent to the fibres. This means that the supersymmetry transformations are best studied in terms of the geometry of vector fields on this bundle in which the projection onto space-time, and hence the splitting into fibres plays a secondary role.

If the projection onto a conventional space-time is removed altogether from the fundamental structures, one obtains the superspace formalism whose local version is used in Wess' lecture. Mathematically it abolishes the asymmetry between the "vertical" direction, along the fibres of an exterior bundle approach in which the exterior algebra appears; and the "horizontal" direction, moving in a purely real manifold. Instead all directions in the basic space have a Grassman algebra character (although this is not made explicit by the formalism in the case of "even" tangent directions which behave algraically as if they were real).

The most natural mathematical basis for this theory is the supermanifold idea, which also appeals to the geometric intuition of many physicists. There are several suggested choices for the topology and the smoothness structure. For one of these versions Batchelor could show the equivalence to the graded manifold approach of Kostant. Several applications and developments of the theory of superspaces including the striking Berezin integration on supermanifolds are presented by Rogers in her lectures.

We conclude that probably the main problem for future collaboration between physicists and mathematicians is the better understanding of the mathematical structures, which one hopes can explain the elementary particle spectrum. One would like to know what is the mathematical basis of setting certain components of the torsion equal to zero in Wess' formulation of superspace. We hope that the conference will act as a start for the solution of such problems and also will begin to shed light on the theoretical and experimental significance of superspace as a spacetime geometry of the microscopic world.

We are most grateful to the Scientific Affairs division of NATO for their generous and efficiently delivered financial support without which the workshop would have been impossible and for their help in the publication of these proceedings. And we are also much indebted to the Hochschule der Bundeswehr Hamburg for providing hospitality for the workshop and for all the participants.

C.J.S. Clarke A. Rosenblum H.-J. Seifert

NON-LINEAR REALIZATION OF SUPERSYMMETRY

Julius Wess

Institut für Theoretische Physik
Universität Karlsruhe
Kaiserstrasse 12
75oo Karlsruhe, W. Germany

Supersymmetry demands a Boson-Fermion symmetry. No such symmetry is present in the phenomenological datas of particle physics at the presently observed energies. In this energy range, supersymmetry could only be realized in a spontaneously broken mode. The uniqueness of the supersymmetry algebra as a Boson-Fermion symmetry and the mathematical beauty of a supersymmetry theory persuades us to try as hard as possible to find a trace of supersymmetry in nature.

The aim of this presentation is to develop a mathematical frame which is most adjusted to a strongly but spontaneously broken symmetry. The supersymmetry partners of boson (fermion) fields are lifted to very high masses and they do not influence the phenomenology at low energies. All what is left from the symmetries are low energy theorems. A pair of goldstino particles couples to the energy momentum tensor with an unknown coupling constant. There is no coupling of the goldstino to the supersymmetry current because all the supersymmetry partners have been chosen to be very heavy.

2. The Akulov-Volkov Field

The formulation of such a theory is best done in terms of non-linearly transforming fields [1] . We first develop this formalism and we start from the supersymmetry algebra.

H.-J. Seifert et al. (eds.), Mathematical Aspects of Superspace, 1–13.

$$\{Q^A_\alpha, \bar{Q}_{\dot{\beta}B}\}_+ = 2\,\sigma^m_{\alpha\dot{\beta}}\,P_m\,\delta^A_B$$

$$\{Q^A_\alpha, Q^B_\beta\}_+ = \{\bar{Q}_{\dot{\alpha}A}, \bar{Q}_{\dot{\beta}B}\}_+ = 0 \tag{1}$$

$$[P_m, Q^A_\alpha]_- = [P_m, \bar{Q}_{\dot{\alpha}A}]_- = 0$$

$$[P_m, P_n]_- = 0\ .$$

The Greek indices $(\alpha, \beta, \ldots, \dot{\alpha}, \dot{\beta}, \ldots)$ run from one to two and denote two component Weyl spinors. The Latin indices $(m, n, \ldots)$ run from one to four and identify Lorentz four vectors. The capital indices $(A, B, \ldots)$ refer to an internal space. They run from 1 to some number $N > 1$. The algebra with $N = 1$ is called supersymmetry algebra, while that with $N > 1$ are called extended supersymmetry algebras. All the notation and conventions can be found in ref. [2].

All the formalism follows from the algebra (1) which may be viewed as a Lie algebra with anticommuting parameters $(\theta^\alpha_A, \bar{\theta}^B_{\dot{\beta}})$. This motivates us to define a corresponding group element:

$$G(x, \theta, \bar{\theta}) = e^{i\{-x^m P_m + \theta^\alpha_A Q^A_\alpha + \bar{\theta}^A_{\dot{\alpha}} \bar{Q}^{\dot{\alpha}}_A\}} \tag{2}$$

It is easy to multiply two group elements using Hausdorff's formula

$$e^A e^B = e^{A+B+\frac{1}{2}[A,B]_- + \ldots} \tag{3}$$

because all higher commutators vanish:

$$G(y, \xi, \bar{\xi}) \cdot G(x, \theta, \bar{\theta}) = \tag{4}$$

$$= G(x^m + y^m + i\theta\sigma^m\bar{\xi} - i\xi\sigma^m\bar{\theta},\quad \theta + \xi, \bar{\theta} + \bar{\xi})\ .$$

Multiplication of group elements induces a motion in the parameter space.

$$\{x, \theta, \bar{\theta}\} \to \{x + i\theta\sigma\bar{\xi} - i\xi\sigma\bar{\theta}, \theta + \xi, \bar{\theta} + \bar{\xi}\} \tag{5}$$

This motion may be generated by the differential operators

$$Q_{\alpha}^{A} = \frac{\partial}{\partial\theta_{A}^{\alpha}} - i\,\sigma_{\alpha\dot{\alpha}}^{m}\,\bar{\theta}^{\dot{\alpha}A}\,\frac{\partial}{\partial x^{m}} \tag{6}$$

$$\bar{Q}_{\dot{\alpha}A} = -\frac{\partial}{\partial\bar{\theta}^{\dot{\alpha}A}} + i\,\theta_{A}^{\alpha}\,\sigma_{\alpha\dot{\alpha}}^{m}\,\frac{\partial}{\partial x^{m}}$$

These differential operators represent the infinitesimal group action:

$$\{Q_{\alpha}^{A},\bar{Q}_{\dot{\alpha}B}\}_{+} = 2i\,\sigma_{\alpha\dot{\alpha}}^{m}\,\delta_{B}^{A}\,\frac{\partial}{\partial x^{m}} \tag{7}$$

$$\{Q_{\alpha}^{A},Q_{\beta}^{B}\}_{+} = \{\bar{Q}_{\dot{\alpha}A},\bar{Q}_{\dot{\beta}B}\}_{+} = 0\ .$$

If we would have interpreted (4) as a right multiplication we would have found that the corresponding motion is generated by the following differential operators:

$$D_{\alpha}^{A} = \frac{\partial}{\partial\theta_{A}^{\alpha}} + i\,\sigma_{\alpha\dot{\alpha}}^{m}\,\bar{\theta}^{\dot{\alpha}A}\,\frac{\partial}{\partial x^{m}} \tag{8}$$

$$\bar{D}_{\dot{\alpha}A} = -\frac{\partial}{\partial\bar{\theta}^{\dot{\alpha}A}} - i\,\theta_{A}^{\alpha}\,\sigma_{\alpha\dot{\alpha}}^{m}\,\frac{\partial}{\partial x^{m}}$$

They satisfy the D-algebra:

$$\{D_{\alpha}^{A},\bar{D}_{\dot{\alpha}B}\}_{+} = -2i\,\sigma_{\alpha\dot{\alpha}}^{m}\,\delta_{B}^{A}\,\frac{\partial}{\partial x^{m}} \tag{8a}$$

$$\{D_{\alpha}^{A},D_{\beta}^{B}\}_{+} = \{\bar{D}_{\dot{\alpha}A},\bar{D}_{\dot{\beta}B}\}_{+} = 0$$

$$\{D_{\alpha}^{A},Q_{\beta}^{B}\}_{+} = \{\bar{D}_{\dot{\alpha}A},Q_{\beta}^{B}\}_{+} = \{D_{\alpha}^{A},\bar{Q}_{\dot{\beta}B}\}_{+} = \{\bar{D}_{\dot{\alpha}A},\bar{Q}_{\dot{\beta}B}\}_{+} = 0$$

The space $\{x^{m},\theta_{A}^{\alpha},\bar{\theta}_{\dot{\alpha}}^{B}\}$ is usually called superspace and equ. (5) represents a supersymmetry transformation in superspace. If we consider a surface in superspace, say $\theta_{A}^{\alpha}(x)$, $\bar{\theta}_{\dot{\alpha}}^{A}(x)$, we can study the motion of this surface under a supersymmetry transformation. To avoid too many θ's we call this surface $\tilde{\lambda}_{A}^{\alpha}(x)$, $\tilde{\bar{\lambda}}_{\dot{\alpha}}^{A}(x)$ and we obtain from equ. (5)

$$\tilde{\lambda}_{A}^{\alpha'}(x') = \tilde{\lambda}_{A}^{\alpha}(x) + \xi_{A}^{\alpha} \tag{9}$$

$$\tilde{\bar{\lambda}}_{\dot{\alpha}}^{A'}(x') = \tilde{\bar{\lambda}}_{\dot{\alpha}}^{A}(x) + \bar{\xi}_{\dot{\alpha}}^{A}$$

$$x'^{m} = x^{m} + i\tilde{\lambda}\sigma^{m}\bar{\xi} - i\xi\sigma^{m}\tilde{\bar{\lambda}}\ .$$

This leads to the following infinitesimal variation of $\tilde{\lambda}$:

$$\delta_\xi \tilde{\lambda}^\alpha_A(x) = \tilde{\lambda}^{\alpha'}_A(x) - \tilde{\lambda}^\alpha_A(x) \tag{10}$$

$$= \xi^\alpha_A - i(\tilde{\lambda}\sigma^m\bar{\xi} - \xi\sigma^m\bar{\tilde{\lambda}}) \frac{\partial}{\partial x^m} \tilde{\lambda}^\alpha_A$$

$$\delta_\xi \bar{\tilde{\lambda}}^A_{\dot{\alpha}}(x) = \bar{\tilde{\lambda}}^{A'}_{\dot{\alpha}}(x) - \bar{\tilde{\lambda}}^A_{\dot{\alpha}}(x)$$

$$= \bar{\xi}^A_{\dot{\alpha}} - i(\tilde{\lambda}\sigma^m\bar{\xi} - \xi\sigma^m\bar{\tilde{\lambda}}) \frac{\partial}{\partial x^m} \bar{\tilde{\lambda}}^A_{\dot{\alpha}}$$

This is the transformation law of the Akulov-Volkov (AV-) field. The commutator of two of such transformations can be computed:

$$(\delta_\eta\delta_\xi - \delta_\xi\delta_\eta)\tilde{\lambda}^\alpha_A = \tag{11}$$

$$= -2i(\eta\sigma^m\bar{\xi} - \xi\sigma^m\bar{\eta}) \frac{\partial}{\partial x^m} \tilde{\lambda}^\alpha_A .$$

The transformation law (10) realizes a supersymmetry transformation on the fields $\tilde{\lambda}^\alpha_A$, $\bar{\tilde{\lambda}}^A_{\dot{\alpha}}$.

Another parametrization of the group element (2) is:

$$G'(x,\theta,\bar{\theta}) = e^{-ix^m P_m}\, e^{i\theta Q}\, e^{i\bar{\theta}\bar{Q}} . \tag{12}$$

Multiplication leads to the following motion in the parameter space:

$$\{x^m,\theta,\bar{\theta}\} \to \{x^m + 2i\theta\sigma^m\bar{\xi}, \theta+\xi, \bar{\theta}+\bar{\xi}\}. \tag{13}$$

The corresponding transformation law of the non-linear realization is:

$$\delta_\xi \lambda^\alpha_A = \xi^\alpha_A - 2i\lambda\sigma^m\bar{\xi} \frac{\partial}{\partial x^m} \lambda^\alpha_A \tag{14}$$

$$\delta_\xi \bar{\lambda}^A_{\dot{\alpha}} = \bar{\xi}^A_{\dot{\alpha}} + 2i\xi\sigma^m\bar{\lambda} \frac{\partial}{\partial x^m} \bar{\lambda}^A_{\dot{\alpha}} .$$

The advantage of this transformation law is that λ transforms into itself and not into $\bar{\lambda}$ as well. The two fields are related:

$$\lambda(x) = \tilde{\lambda}(y) \tag{15}$$
$$y^m = x^m - i\tilde{\lambda}(y)\sigma^m \bar{\tilde{\lambda}}(y) .$$

3. Superfields

Linear representations of the supersymmetry algebra (1) are easy to define in terms of superfields. Superfields are functions of the superspace variable $\{x^m, \theta^{\alpha}_{A}, \bar{\theta}^{A}_{\dot{\alpha}}\}$ which should be understood in terms of their power series expansion in θ and $\bar{\theta}$

$$F(x,\theta,\bar{\theta}) = f(x) + \theta\varphi(x) + \bar{\theta}\bar{\chi}(x) + \dots + \theta^{2N}\bar{\theta}^{2N} d(x). \tag{16}$$

All higher powers of $\theta,\bar{\theta}$ vanish. The transformation law of superfields is defined as follows:

$$\begin{aligned}\delta_{\xi} F(x,\theta,\bar{\theta}) &\equiv \delta_{\xi} f(x) + \theta\delta_{\xi}\varphi + \bar{\theta}\delta_{\xi}\bar{\chi} + \\ &+ \dots + \\ &+ \theta^{2N}\bar{\theta}^{2N}\delta_{\xi} d(x) = \\ &= (\xi Q + \bar{\xi}\bar{Q})F(x,\theta,\bar{\theta}),\end{aligned} \tag{17}$$

where Q and $\bar{Q}$ are the differential operators (6). The transformation laws of the component fields $(f,\varphi,\chi,\dots,d)$ may be found from (17) by comparing appropriate powers of $\theta,\bar{\theta}$. The commutator of these transformations satisfies the algebra (1) as a consequence of (7).

Linear combination of superfields, products of superfields, space-time derivatives of superfields and $D,\bar{D}$ derivatives of superfields are again superfields.

The definition of the transformation law (17) has as a consequence that the highest component of a superfield will always transform into a space time derivative. Integrating a highest component of a superfield over d^4x yields an invariant. This way all the supersymmetric Lagrangians may be formulated.

It is always possible to construct a superfield from any realization of the algebra (1). We start from an object, say $a(x)$, that transforms under supersymmetry transformations such that

$$(\delta_{\eta}\delta_{\xi} - \delta_{\xi}\delta_{\eta})a = -2i(\eta\sigma^m\bar{\xi} - \xi\sigma^m\bar{\eta})\frac{\partial}{\partial x^m} a . \tag{18}$$

We apply the operator $\exp[\delta_\theta]$ to a, its action is defined because δ_θ a is supposed to be known as well as the transformation law of all the other fields into which a transforms.

$$A(x,\theta,\bar{\theta}) = e^{\delta_\theta} a \; . \tag{19}$$

This is a superfield, the transformation law (17) is a consequence of the algebraic relation (18) for a and all its partners in the realization.

Using Hausdorff's formula it is also possible to derive the following relation:

$$(\xi^\alpha_A D^A_\alpha + \bar{\xi}^A_{\dot{\alpha}} \bar{D}^{\dot{\alpha}}_A)\, e^{\delta_\theta} = e^{\delta_\theta}\, \delta_\xi \; . \tag{20}$$

This is a relation which we shall use frequently to derive constraint equations on superfields.

Applying this construction to the Akulov-Volkov-field, the transformation law of which is given through (9) or (14), we arrive at a superfield which we shall call Akulov-Volkov superfield $\tilde{\Lambda}^A_\alpha$ or Λ^A_α [3].

The higher $\theta,\bar{\theta}$ components of $\tilde{\Lambda},\Lambda$ are not new degrees of freedom, they are functions of the lower component and its space time derivative. These functions are arranged in such a way that the whole object transforms linearly, like a superfield.

From (20) follows that the AV-superfield satisfies the following constraint equations:

$$D_{\beta B}\, \tilde{\Lambda}^A_\alpha = \varepsilon_{\alpha\beta}\, \delta^A_B + i\, \sigma^m_{\beta\dot{\rho}}\, \bar{\tilde{\Lambda}}^{\dot{\rho}}_B \frac{\partial}{\partial x^m} \tilde{\Lambda}^A_\alpha \tag{21}$$

$$\bar{D}^B_{\dot{\beta}}\, \tilde{\Lambda}^A_\alpha = -\, i\, \tilde{\Lambda}^{\rho B}\, \sigma^m_{\rho\dot{\beta}} \frac{\partial}{\partial x^m} \tilde{\Lambda}^A_\alpha$$

or

$$D_{\beta B}\, \Lambda^A_\alpha = \delta^A_B\, \varepsilon_{\alpha\beta} \tag{22}$$

$$\bar{D}^B_{\dot{\beta}}\, \Lambda^A_\alpha = -\, 2i\, \Lambda^{\rho B}\, \sigma^m_{\rho\dot{\beta}} \frac{\partial}{\partial x^m} \Lambda^A_\alpha \; .$$

We would have defined the AV-superfield through equs. (21) or (22). A solution of those equations could have rendered a superfield whose components are exactly the same function of its lowest component as was found from the previous construction of $\tilde{\Lambda}^A_\alpha$ or Λ^A_α .

4. Standard Fields

Any representation of the Poincaré group can be extended to a realization of the supersymmetry algebra (1) with the help of the Akulov-Volkov-field [4]. Starting from any tensor field $C_{Ind}(x)$ where Ind refers to any indices we define

$$\delta_\xi \tilde{C}_{Ind} = - i(\tilde{\lambda}\sigma^m\bar{\xi}-\xi\sigma^m\bar{\tilde{\lambda}}) \frac{\partial}{\partial x^m} \tilde{C}_{Ind} \qquad (23)$$

or

$$\delta_\xi C_{Ind} = - 2i \lambda\sigma^m\bar{\xi} \frac{\partial}{\partial x^m} C_{Ind} . \qquad (24)$$

It can be verified by a direct calculation that this transformation law realizes the supersymmetry algebra:

$$(\delta_\eta\delta_\xi-\delta_\xi\delta_\eta)\tilde{C}_{Ind} = - 2i(\eta\sigma^m\bar{\xi}-\xi\sigma^m\bar{\eta}) \frac{\partial}{\partial x^m} \tilde{C}_{Ind} \qquad (25)$$

$$(\delta_\eta\delta_\xi-\delta_\xi\delta_\eta)C_{Ind} = - 2i(\eta\sigma^m\bar{\xi}-\xi\sigma^m\bar{\eta}) \frac{\partial}{\partial x^m} C_{Ind} .$$

We can associate a superfield with both fields following the construction outlined before. From equ. (20) we can derive the constraint equs.

$$D_{\beta A} \tilde{C}_{Ind} = i \sigma^m_{\beta\dot{\rho}} \bar{\tilde{\Lambda}}^{\dot{\rho}}_A \frac{\partial}{\partial x^m} \tilde{C}_{Ind} \qquad (26)$$

$$\bar{D}^A_{\dot{\beta}} \tilde{C}_{Ind} = - i \tilde{\Lambda}^{\rho A} \sigma^m_{\rho\dot{\beta}} \frac{\partial}{\partial x^m} \tilde{C}_{Ind}$$

or

$$D_{\beta A} C_{Ind} = 0 \qquad (27)$$

$$\bar{D}^A_{\dot{\beta}} C_{Ind} = - 2i \Lambda^\rho\sigma^m_{\rho\dot{\beta}} \frac{\partial}{\partial x^m} C_{Ind}.$$

We would like to show that any realization of the supersymmetry algebra may be decomposed into objects that transform as standard matter fields [5]. We start by decomposing a superfield $F(x,\theta,\bar{\theta})$. Its transformation law was given in equ. (17). It can be written in the following form:

$$\delta_\xi F(x,\theta,\bar{\theta}) = F(x+i\theta\sigma^m\bar{\xi}-i\xi\sigma^m\bar{\theta},\xi+\theta,\bar{\xi}+\bar{\theta}) - F(x,\theta,\bar{\theta}). \qquad (28)$$

If we replace $\theta,\bar\theta$ by $-\lambda$, $-\bar{\tilde\lambda}$ we obtain a combination of the component fields of F and the AV-field

$$\tilde f = F(x,-\tilde\lambda,-\bar{\tilde\lambda}) = f(x) - \tilde\lambda\varphi - \bar{\tilde\lambda}\bar\chi + \ldots + \tag{29}$$
$$+ \tilde\lambda^{2N}\bar{\tilde\lambda}^{2N} d(x).$$

The transformation property of $\tilde f$ can be computed from (28) and (10). We obtain:

$$\delta_\xi \tilde f(x) = F(x-i\tilde\lambda\sigma^m\bar\xi+i\xi\sigma^m\bar{\tilde\lambda},-\tilde\lambda+\xi-\delta\tilde\lambda,-\bar{\tilde\lambda}+\bar\xi-\delta\bar{\tilde\lambda}) - \tag{30}$$
$$- F(x,-\tilde\lambda,-\bar{\tilde\lambda})$$
$$= - i(\tilde\lambda\sigma^m\bar\xi-\xi\sigma^m\bar{\tilde\lambda}) \frac{\partial}{\partial x^m} \tilde f(x).$$

This is exactly the transformation law (23).

Instead of F we could have replaced $\theta,\bar\theta$ by $-\lambda$, $-\bar{\tilde\lambda}$ in the superfield $D_\alpha F(x,\theta,\bar\theta)$ and we would have obtained a standard form

$$\tilde\varphi_\alpha = \varphi_\alpha + \ldots \tag{31}$$

Each component of F can be made to a lowest component of a superfield by applying appropriate powers of $D,\bar D$ to F. Replacing in this superfield $\theta,\bar\theta$ by $-\tilde\lambda$, $-\bar{\tilde\lambda}$ yields a standard field as a function of the original component fields and the AV field. All fields obtained that way amount to an allowed reparametrization [6] of the component field of F. We have decomposed F into spectator fields.

Before we have shown that any realization of the supersymmetry algebra may be put into the form of a superfield. Now we have shown that a superfield can be decomposed into standard fields. Therefore, we can reparametrize any realization in terms of standard fields.

Let me finally indicate how the inverse transformation can be constructed. We want to obtain a superfield

$$F(x,\theta,\bar\theta) = f(x) + \theta\varphi + \bar\theta\bar\chi + \ldots \ldots + \theta^{2N}\bar\theta^{2N}d \tag{32}$$

whose components transform linearly. We have standard fields $\tilde f,\tilde\varphi,\tilde\chi, \ldots \tilde d$ at our disposal. From these standard fields we construct standard superfields according to (19). We call them $\tilde{F},\tilde{\phi},\tilde{\chi}\ldots\tilde{D}$.

These superfields we combine as follows:

$$F = \tilde{F} + \Lambda\tilde{\phi} + \bar{\Lambda}\tilde{\bar{\chi}} + \ldots + \Lambda^{2N}\bar{\Lambda}^{2N}\tilde{D} \;. \tag{33}$$

This is a superfield. If we would have derived $\tilde{f},\tilde{\varphi},\tilde{\chi}\ldots\tilde{d}$ from a superfield F as was done before and then construct F through equ. (33) we would have obtained an identity. Equ. (33) shows how each superfield can be reconstructed from its spectator fields.

5. N > 1 / N = 1

In this chapter I will address the problem of non-linear realizations of an N > 1 supersymmetry such that an N = 1 supersymmetry is realized linearly. We [7] shall study surfaces in superspace which are functions of x^m and θ^α, $\bar{\theta}_{\dot{\alpha}}$ as well. For convenience, we introduce a notation where θ^α, and $\bar{\theta}_{\dot{\alpha}}$ are denoted by $\theta^\alpha, \bar{\theta}_{\dot{\alpha}}$ and $\theta^\alpha_2 \ldots \theta^\alpha_N$, $\bar{\theta}^2_{\dot{\alpha}} \ldots \bar{\theta}^N_{\dot{\alpha}}$ by θ'^{α}_A, $\bar{\theta}'^{A}_{\dot{\alpha}}$. The index runs from 1 to N - 1.

The transformation (5) in superspace takes the following form in this notation:

$$x^m \to x^m + i\theta\sigma^m\bar{\xi} - i\xi\sigma^m\bar{\theta} + i\theta'^{\alpha}_A\sigma^m_{\alpha\dot{\alpha}}\bar{\xi}^{\dot{\alpha}\,'A} - i\xi'^{\alpha}_A\sigma^m_{\alpha\dot{\alpha}}\bar{\theta}^{\,'A\dot{\alpha}}$$

$$= x^m + i\theta\sigma^m\bar{\xi} - i\xi\sigma^m\bar{\theta} + i\theta'\sigma^m\bar{\xi}' - i\xi'\sigma^m\bar{\theta}'$$

$$\theta^\alpha \to \theta^\alpha + \xi^\alpha$$

$$\bar{\theta}_{\dot{\alpha}} \to \bar{\theta}_{\dot{\alpha}} + \bar{\xi}_{\dot{\alpha}}$$

$$\theta'^{\alpha}_A \to \theta'^{\alpha}_A + \xi'^{\alpha}_A$$

$$\bar{\theta}'^{A}_{\dot{\alpha}} \to \bar{\theta}'^{A}_{\alpha} + \bar{\xi}'^{A}_{\dot{\alpha}} \tag{34}$$

The transformation law of a surface $\tilde{\chi}^A_\alpha(x^m,\theta,\bar{\theta})$ can now be deduced from (34). We list the transformation under ξ and ξ' respectively.

$$\begin{aligned}\delta_\xi\tilde{\chi}^A_\alpha &= \tilde{\chi}^A_\alpha(x^m + i\theta\sigma^m\bar{\xi} - i\xi\sigma^m\bar{\theta}, \theta^\alpha + \xi^\alpha, \bar{\theta}_{\dot{\alpha}} + \bar{\xi}_{\dot{\alpha}}) \\ &\quad - \tilde{\chi}^A_\alpha(x^m, \theta^\alpha, \bar{\theta}_{\dot{\alpha}}) \\ &= (\xi^\alpha Q_\alpha + \bar{\xi}_{\dot{\alpha}}\bar{Q}^{\dot{\alpha}})\chi^A_\alpha \;.\end{aligned} \tag{35}$$

We conclude that $\tilde{\chi}^A_\alpha$ transforms like an N = 1 superfield under the ξ transformation. Under the ξ' transformation we find

$$\delta_{\xi'}\tilde{\chi}^A_\alpha = \tilde{\chi}'^A_\alpha(x,\theta,\bar{\theta}) - \tilde{\chi}^A_\alpha(x,\theta,\bar{\theta}) \qquad (36)$$

with

$$\tilde{\chi}'^A_\alpha(x',\theta,\bar{\theta}) = \xi'^A_\alpha + \tilde{\chi}^A_\alpha(x,\theta,\bar{\theta})$$
$$x'^m = x^m + i\tilde{\chi}\sigma^m\bar{\xi}' - \xi'\sigma^m\bar{\tilde{\chi}} \ . \qquad (37)$$

This leads to the following non-linear transformation law:

$$\delta_{\xi'}\tilde{\chi}^A_\alpha = \xi'^A_\alpha - i(\tilde{\chi}\sigma^m\bar{\xi}' - \xi'\sigma^m\bar{\tilde{\chi}})\frac{\partial}{\partial x^m}\tilde{\chi}^A_\alpha \qquad (38)$$
$$\delta_{\xi'}\bar{\tilde{\chi}}^{\dot{\alpha}} = \bar{\xi}'^{\dot{\alpha}}_A - i(\tilde{\chi}\sigma^m\bar{\xi}' - \xi'\sigma^m\bar{\tilde{\chi}})\frac{\partial}{\partial x^m}\bar{\tilde{\chi}}^{\dot{\alpha}}_A$$

If we compute this with equ. (10) we find that $\tilde{\chi}$ transforms like a AV-Field for the N ≠ 1 supersymmetry transformations. That (35) and (36) really realize the supersymmetry algebra can be demonstrated by computing the commutator of these transformations.

It is obvious how to generalize (14) to the case N > 1 / N = 1. We replace λ^α_A in (14) by an N = 1 superfield χ^α_A and have A run from 1 to N - 1.

$$\delta_{\xi'}\chi^\alpha_A = \xi'^\alpha_A - 2i\chi\sigma^m\bar{\xi}'\frac{\partial}{\partial x^m}\chi^\alpha_A \qquad (39)$$
$$\delta_{\xi'}\bar{\chi}^A_{\dot{\alpha}} = \bar{\xi}'^A_{\dot{\alpha}} - 2i\xi'\sigma^m\bar{\chi}\frac{\partial}{\partial x^m}\tilde{\chi}^A_{\dot{\alpha}}$$

This parametrization of χ is related to $\tilde{\chi}$ as was λ to $\tilde{\lambda}$ in equ. (15). The transformation law (34) has the advantage that χ might be chosen to be N = 1 chiral, or antichiral:

$$\bar{D}_{\dot{\alpha}}\chi^\alpha_A = 0 \qquad : \text{chiral}$$
$$\qquad\qquad\qquad\qquad (40)$$
$$D_\alpha \hat{\chi}^\beta_A = 0 \qquad : \text{antichiral}$$

The transformation law of standard fields (23) and (24) can be easily generalized:

$$\delta\tilde{C}_{Ind} = -i(\tilde{\chi}\sigma^m\bar{\xi}' - \xi'\sigma^m\bar{\tilde{\chi}})\frac{\partial}{\partial x^m}\tilde{C}_{Ind} \qquad (41)$$

or

$$\delta C_{Ind} = - 2i\chi\sigma^m\bar{\xi}' \frac{\partial}{\partial x^m} C_{Ind} \tag{42}$$

or

$$\delta \hat{C}_{Ind} = - 2i\hat{\chi}\sigma^m\bar{\xi}' \frac{\partial}{\partial x^m} \hat{C}_{Ind}. \tag{43}$$

In all three cases, C_{Ind} is supposed to be an N = 1 superfield. In the first case (41), $\tilde{C}_{Ind}$ can be constrained to be hermitean, in the second case, (42), to be chiral and in the third case, (43), to be antichiral. The transformation laws (41), (42), (43) would leave such a constraint invariant.

All these fields can be lifted to superfields exactly in the same fashion as in the previous chapter and they might be defined with constraint equations as well.

6. N = 1 Supergravity

The N = 1 supergravity transformations can be realized in a "chiral" superspace [8]. We follow the notation of ref. [2], chapter XX, where the supergravity transformation law of a chiral superfield was derived in the form

$$\delta\phi = - (\eta^m \frac{\partial}{\partial x^m} + \eta^\alpha \frac{\partial}{\partial \Theta^\alpha})\phi \tag{44}$$

$$\phi = A + \sqrt{2}\, \Theta^\alpha \chi_\alpha + \theta^\alpha \theta_\alpha F$$

$$\eta^m = 2i\Theta\sigma^m\bar{\xi}(x) + \Theta\Theta\bar{\psi}_n \bar{\sigma}^m\sigma^n\bar{\xi}(x)$$

$$\begin{aligned}\eta^\alpha = {} & \xi^\alpha(x) - i\Theta\sigma^m\bar{\xi}(x)\psi_m^\alpha \\ & + \Theta\Theta\{\frac{1}{3} M^*\xi^\alpha(x) + \frac{1}{6} b_a(\varepsilon\sigma^a\bar{\xi})^\alpha \\ & \quad - i\,\omega_m^{\alpha\beta}(\sigma^m\bar{\xi}(x))_\beta \\ & \quad - \frac{1}{2}\psi_n^\alpha(\bar{\psi}_m\bar{\sigma}^n\sigma^m\bar{\xi})\}.\end{aligned}$$

We rewrite this transformation law in the form

$$\delta\phi = \phi(x+\eta^m, \Theta^\alpha+\eta^\alpha) - \phi(x,\Theta) \tag{44a}$$

and we conclude that supergravity induces the following motion in our new superspace:

$$\delta x^m = \eta^m(x,\Theta) \tag{45}$$

$$\delta\Theta^\alpha = \eta^\alpha(x,\Theta)$$

We consider a surface $\Theta^\alpha(x)$ and derive how it changes under the coordinate transformations (45):

$$\delta\lambda^\alpha(x) = \eta^\alpha(x,\lambda) - \eta^m(x,\lambda)\frac{\partial}{\partial x^m}\lambda^\alpha(x) \tag{46}$$

This transformation law should be compared with equ. (3.12) in ref. [3]. For this purpose we write (46) more explicitly:

$$\begin{aligned}\delta\lambda^\alpha(x) = {} & \xi^\alpha - i\lambda\sigma^m\bar{\xi}\psi_m^\alpha + \lambda^2\{\tfrac{1}{3}M^*\xi^\alpha + \tfrac{1}{6}b_a(\varepsilon\sigma^a\bar{\xi})^\alpha \\ & - i\omega_m^{\alpha\beta}(\sigma^m\bar{\xi})_\beta - \tfrac{1}{2}\psi_n^\alpha(\bar{\psi}_m\bar{\sigma}^n\sigma^m\bar{\xi})\} \\ & - (2i\lambda\sigma^m\bar{\xi} + \lambda^2\bar{\psi}_n\bar{\sigma}^m\sigma^n\bar{\xi})\frac{\partial}{\partial x^m}\lambda^\alpha(x)\end{aligned} \tag{47}$$

This is exactly the transformation law mentioned above if we replace λ by $-\lambda$.

REFERENCES

[1] D.V. Volkov and V.P. Akulov, JETP Lett.16, 438(1972);
V.P. Akulov and D.V. Volkov, Phys.Lett. 46B (1973)1o9;
D.V. Volkov and V. Soraka, JETP Lett. 18 (1973) 312.

[2] J. Wess and J. Bagger, Supersymmetry and Supergravity, Princeton Series in Physics, Princeton University Press, USA.

[3] S. Samuel and J. Wess, Nucl. Phys. B221 (1983) 53.

[4] S. Samuel and J. Wess, Nucl.Phys. B221 (1983) 53;
J. Wess, Lectures given at Dubrovnik, Karlsruhe preprint (1983).

[5] E. Ivanov and A. Kapustnikov, J. Phys. A 11 (1978) 58, J. Phys.G 8 (1982) 167.

[6] S. Coleman, J. Wess and B. Zumino, Phys. Rev. 177 (1969) 2239;
C. Callan, S. Coleman, J. Wess and B. Zumino, Phys. Rev. 177 (1969) 2247;

J. Wess, Springer Tracts in Modern Physics, Vol. 50 (1969) 132.

[7] J. Bagger and J. Wess, SLAC-PUB-3255 (1983).

[8] J. Wess and B. Zumino, Phys. Lett. 74B (1978) 51.

FIELDS, FIBRE BUNDLES AND GAUGE GROUPS

C.J.S. Clarke

Department of Mathematics, University of York

Abstract A survey of the main geometrical concepts underlying modern gauge theories in theoretical physics.

1. MANIFOLDS

These are ubiquitous in mathematical physics; examples are

(i) space-time

(ii) the space of all possible fields in a classical field theory

(iii) the space of all (smooth) paths starting from a given point, as used for path-integral theories

and so on. Examples (ii) and (iii) are not finite-dimensional, but all are topological spaces - so that it makes sense to say that a sequence of elements tends to a given limit, and one can talk about neighbourhoods - and all have the property that every element has a neighbourhood that is homeomorphic to a neighbourhood in some fixed real vector space V. The manifold is then said to be modelled on V. More formally, one requires the existence of an atlas consisting of a collection of charts

H.-J. Seifert et al. (eds.), Mathematical Aspects of Superspace, 15–45.

(U_i, ψ_i), where i runs over some indexing set I, with U_i an open set in the manifold M, the set $(U_i)_{i\in I}$ covering M, and $\psi_i : U_i \to V$ a homeomorphism of U_i into V.

In the case where M is space-time, then $V = \mathbb{R}^4$ and for every event x in a given U_i the point $\psi_i(x)$ is the quadruple $(\psi_i(x)^1, \psi_i(x)^2, \psi_i(x)^3, \psi_i(x)^4)$ of <u>coordinates</u> for x (written briefly as (x^1, x^2, x^3, x^4)).

Suppose $U_i \cap U_j \neq \emptyset$, Define $\chi_{ij} = \psi_j \circ (\psi_i | U_i \cap U_j)^{-1}$, i.e.

$$\chi_{ij} : \psi_i(U_i \cap U_j) \to \psi_j(U_i \cap U_j) \subset V.$$

So χ_{ij} is the map taking coordinates with respect to ψ_i into the coordinates with respect to ψ_j. In all the manifolds encountered in physics there is some restriction on the maps χ_{ij}: they are required to lie in some special subset Γ of maps of open sets of V into V. A minimal requirement is that the maps in Γ are all differentiable, and may even be C^∞, so that a function on M that is smooth in one coordinate representation is smooth in all. (When V is infinite-dimensional this involves complications in choosing the appropriate definition of differentiability with respect to a norm on V).

For consistency, we require that if the composition of two maps in Γ is defined, then the result is also in Γ. This is expressed by saying that Γ is a <u>pseudogroup</u> (1).

If the model-space V has some additional structure then we can require that the maps in Γ preserve that structure. For example, if we believe that analyticity is important in physics, we may work with theories in which $V = \mathbb{C}^n$ and the coordinate-transformation maps χ_{ij} are holomorphic functions from neighbourhoods of $\mathbb{C}^n$ into $\mathbb{C}^n$.

The example central to this workshop is the case where V has a structure based on a Grassman algebra (a pseudo-Euclidean space) and Γ is the pseudogroup of transformations satisfying a generalisation of differentiability with respect to this structure. The resulting manifold is then a supermanifold (see the lectures by Batchelor and Rogers in this volume).

2. FIBRE BUNDLES

2.1 Fields

From now on M will denote space-time. In the notation of §1, $V = \mathbb{R}^4$ and Γ comprises all C^∞ maps of open sets of $\mathbb{R}^4$ into $\mathbb{R}^4$. (M also has to satisfy certain topological requirements, e.g. it must be connected, as there is only one universe.)

The simplest (classical) field is a real scalar field, i.e. a real valued function ϕ on M. Normally one requires that each coordinate representation

$$\phi_i \equiv \phi \circ \psi_i^{-1} \;:\; \mathbb{R}^4 \supset \psi_i(U_i) \to \mathbb{R}$$

be smooth, so that it makes sense to differentiate ϕ_i in order to apply field equations. More generally, a vector field $\underset{\sim}{\phi}$ could be an $\mathbb{R}^n$-valued function on M, or, more general still, a function taking values in some abstract vector space F. In this case the space F is a global object of the theory, not linked in any way to the space-time structure.

However, in general relativity the commonest situation is to have a field ϕ whose value $\phi(p)$ at p is a vector or a tensor at p (e.g. the electromagnetic tensor $F_{\mu\nu}$). In this case the field at p takes values in a vector space F_p, say, which is defined locally at each point p. This is the situation not only in relativity but in all gauge theories, and so by now it can be regarded as the "normal situation" in physical theories. The union $X = \bigcup_{p\in M} F_p$ of all the target-spaces for the field is called a fibre bundle; each F_p is a <u>fibre</u>.

We now proceed to the formal definition, which includes the idea of the group G of the bundle: the "local symmetry group" of a physical theory. Examples will follow.

2.2 Coordinate bundles

The basic idea is that, just as a manifold looks locally like a vector space V, so a fibre bundle looks locally like the situation where each F_p is the same global F. In that case we could construct X as the Cartesian product MxF and define $F_p = \{p\} \times F$. (Such an X is called a product bundle.) To define a general fibre bundle we relax this, requiring that in a neighbourhood W_i of M the set $X_i = \bigcup_{p \in W_i} F_p$ should be homeomorphic to $W_i \times F$ by a map $\theta_i : X_i \to W_i \times F$; but the choice of θ_i may be arbitrary, and there need exist no map that covers the whole of M. Two different maps θ_i and θ_j will disagree in general, so that

$$\theta_i(v) = (p, f_1), \ \theta_j(v) = (p, f_2)$$

for $v \in F_p$, with $f_1 \neq f_2$. We require, however, that f_1 and f_2 be related by a symmetry transformation: an action of a group G on F.

This leads us to define a <u>coordinate bundle</u> (2) to be the collection $(X, \pi, M, F, G, \{W_i\}_{i \in I}, \{\theta_i\}_{i \in I})$ where

(i) $\pi : X \to M$ is a continuous map of topological spaces;

(ii) G is a topological group which acts continuously and effectively on the topological space F on the left;

(iii) $\{W_i\}_{i \in I}$ is an open cover of M;

(iv) for each $i \in I$, $\theta_i : \pi^{-1}(W_i) \to W_i \times F$ is a homeomorphism satisfying $\pi \circ \theta_i^{-1}(p, f) = p$;

(v) for all $i, j \in I$ with $W_i \cap W_j \neq \emptyset$ there is a continuous map $\gamma_{ij} : W_i \cap W_j \to G$ such that for every $w \in W_i \cap W_j$ and every $f \in F$ we have $\theta_j \circ \theta_i^{-1}(w, f) = (w, \gamma_{ij}(w) f)$.

We write $F_p = \pi^{-1}(p)$, $X_i = \pi^{-1}(W_i)$. In all the cases we shall meet, everything in sight will be smooth and G will be a Lie group acting on the smooth manifold F. We call F the standard fibre and π the projection. The maps θ_i are called trivialisations or coordinate maps.

2.3 Fibre bundles

Two coordinate bundles with the same X,π,M,F and G but different covers $\{W_i\}_{i\in I}$, $\{W_i'\}_{i\in I}$, and maps $\{\theta_i\}_{i\in I}$, $\{\theta_i'\}_{i\in I}$, are called equivalent if $(X,\pi,M,F,G,\{W_i\}\cup\{W_i'\},\{\theta_i\}\cup\{\theta_i'\})$ is a coordinate bundle. Then we define a fibre bundle as an equivalence class of coordinate bundles, denoting it by (X,π,M,F,G).

2.4 Examples

The fundamental example for relativity theory is the tangent bundle TM to space-time. Here $F = \mathbb{R}^4$, $G = GL(4,\mathbb{R})$, and for any atlas (U_i,ψ_i) for M we have a coordinate bundle defined by taking $W_i = U_i$, $\theta_i(v) = (x;v^1,v^2,v^3,v^4)$, where v^α are the components of the tangent vector v at x in the coordinates ψ_i. The map γ_{ij} has a value at each point p given by the coordinate-transformation matrix $[\partial\psi_j{}^\alpha/\partial\psi_i{}^\beta]$.

To see how the global topology of a tangent bundle may be non-trivial, consider the low-dimensional analogue in which instead of space-time we take M to be a 2-sphere S^2. This can be covered by two charts, defined by stereographic projection from the North and South poles, related by

$$x' = -x/(x^2+y^2),\quad y' = y/(x^2+y^2) \tag{1}$$

where (x,y) and (x',y') are the components of $\psi_i(p)$ in the two charts $(i = 1,2)$. Over each chart the tangent bundle is represented by $\theta_i(X_i) = \mathbb{R}^2 \times \mathbb{R}^2$; one chart covers S^2 - {North Pole}, the other S^2 - {South Pole}.

We can think of the entire tangent bundle as derived by identifying the part of $\theta_1(X_1)$ not over the origin with the corresponding part of $\theta_2(X_2)$, using the transformation matrix

$$\gamma_{12} = \partial(x',y')/\partial(x,y) = \frac{1}{r}\begin{bmatrix}\cos 2\theta & \sin 2\theta\\ -\sin 2\theta & \cos 2\theta\end{bmatrix}$$

where $x = r\cos\theta$, $y = r\sin\theta$. We see that the transformation of components of tangent vectors between X_1 and X_2 involves a rotation of the $\mathbb{R}^2$ by an angle 2θ that depends on the position in M. Thus TM is derived by taking two copies of $\mathbb{R}^2 \times \mathbb{R}^2$ and identifying them by using (1) on the first $\mathbb{R}^2$ so as to form the sphere, while identifying the second $\mathbb{R}^2$ with a twist. (see figure 1).

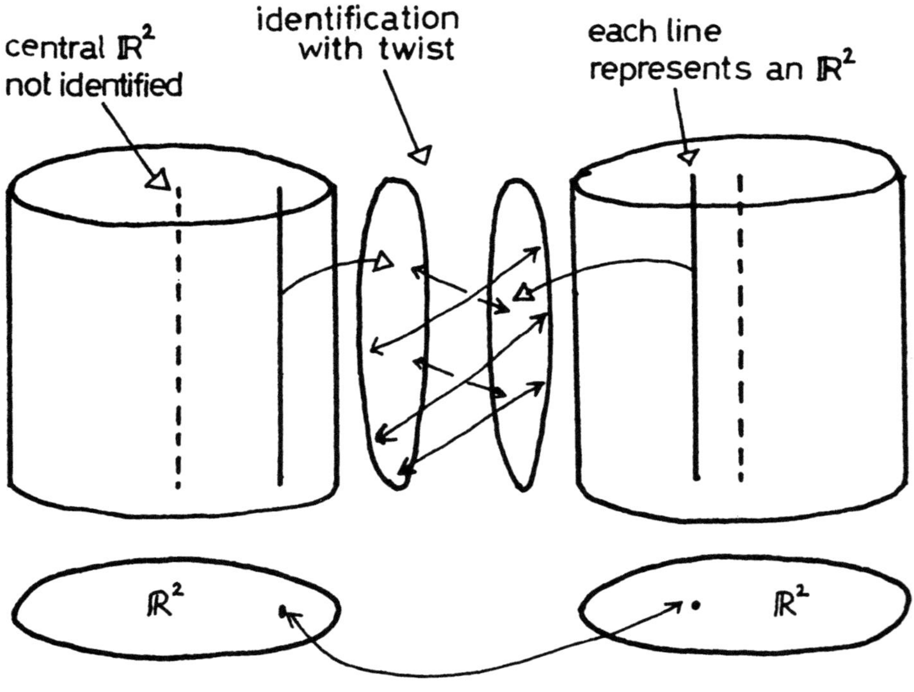

It can be shown without much difficulty that there is no way of removing this "twist", in the sense that there is no globally definable $\theta : X \to M \times \mathbb{R}^2$.

This situation is closely analogous to the phenomenon of "topological charge" in Yang-Mills theories, where boundary conditions are imposed that lead to the possibility of compactifying space-time into an S^4 and the field takes values in a bundle that has a "twist" of precisely the above form.

The simplest example is provided by electrodynamics for a complex scalar field. Here the group $G = U(1)$ and $F = \mathbb{C}$ (with the standard action of G). For a given choice of θ over a neighbourhood W in M, $\theta(\pi^{-1}W) = W\times\mathbb{C}$.

A field ϕ is a function $M \to X$, which can be represented over W via θ as the function

$$\theta\circ\phi \;:\; W \to W\times\mathbb{C}$$
$$p \mapsto (p,\psi(p))$$

say, with ψ a complex-valued function on W. Different choices of θ (different choices of gauge) give rise to different functions ψ, any two such functions being related by

$$\psi'(p) = \gamma(p)\psi(p)$$

where $\gamma(p) \in U(1)$; in other words, the usual gauge transformation of multiplication by a complex number of modulus unity.

2.5 Fields and geometry

For practical calculations one works locally, fixing a particular gauge θ and regarding the fields ϕ as taking values in a fixed F by writing $\theta\circ\phi(p) = (p,\psi(p))$, $\psi(p)\in F$, as in the example above. What, then, is the advantage of talking about fibre bundles? why not just deal with fields ψ taking values in a fixed F, and afterwards impose gauge-invariance?

There are three advantages. First, there may be important global information, such as the topological charge in Yang-Mills theories, that cannot be seen (or can only be seen indirectly) when one works purely locally. Secondly, there are many geometrical structures related to curvature that can be defined directly on a fibre bundle, without using particular trivialisations θ. Such structures will, when expressed in terms of a θ, automatically be gauge-invariant. So the geometrical approach guarantees explicit gauge-independence at every step. Third, the use of geometrical concepts such as curvature allows one to take over all the machinery of classical differential geometry to prove results that would be extremely non-obvious when expressed in terms of a particular gauge (θ).

The use of geometrical language highlights a problem which is disguised in a gauge-dependent approach using coordinates. Namely, we cannot take it for granted that the fibre bundle is simply a passive arena in which fields evolve according to some dynamics. In general relativity the manifold of space-time is not given a priori but is determined dynamically along with the metric field; and it seems likely that in the case of topological charge the global structure of the fibre bundle should again be regarded as a dynamical variable. So we are not dealing with the dynamics of a field in a bundle, but with the dynamics of a bundle-with-fields. It is not clear how this should best be approached. One line of attack, for example, is to enlarge the bundle until it becomes necessarily trivial, and, to compensate, impose constraints on the fields. This is a dominant approach in supergravity, but the enlargement of the bundle tends to hide the geometrical structures, which become encoded in the constraints.

2.6 Principal bundles

In the case where $F = G$ and the action of G on itself is left multiplication we call the resulting fibre bundle a principal bundle, denoted by the quadruple (X,π,M,G). Some physicists call this a group manifold. Whereas in a fibre bundle G acts on the standard fibre (conventionally, on the left) to provide coordinate transformations, in a principal bundle G also acts on X itself, but on the right. To define this action, let $q \in \pi^{-1}(x)$ with $x \in W_i \subset M$ and suppose that $\theta_i(q) = (x,f)$ for $f \in G(=F)$. For any $h \in G$ define $qh = \theta_i^{-1}(x,fh)$. Then this definition is independent of the choice of coordinate (gauge) θ, because changing θ multiplies f by an element of G on the left, and this commutes with the multiplication by h on the right.

Because right-multiplication is a simply transitive action on G, and each fibre is isomorphic to G, it follows that the action of G on X is simply transitive on each fibre; i.e. given q_1 and q_2 in a fibre there is a unique h in G such that $q_1 h = q_2$.

The standard example of a principal bundle in relativity theory is the frame bundle LM, consisting of all quadruples (E_1, E_2, E_3, E_4) of linearly independent vectors at all points of M. The projection π takes $(E_\alpha)\alpha=1,\ldots,4$ to the point p in M at which the E_α are situated, while every chart (W_i, ψ_i) of M defines a map from $\pi^{-1}(W_i)$ to $W_i \times GL(4,\mathbb{R})$ by

$$\underset{\sim}{E} = (E_\alpha)_\alpha \to (\pi(\underset{\sim}{E}), A(\underset{\sim}{E})),\ A(\underset{\sim}{E})^\beta{}_\alpha = E_\alpha^\beta$$

where E_α^β is the β'th component of E_α in the coordinates ψ_i. The group $GL(4,\mathbb{R})$ acts on the coordinates $A(\underset{\sim}{E})$ by coordinate transformations on the left of the form

$$A(\underset{\sim}{E}) \mapsto PA(\underset{\sim}{E}) \quad (P \in GL(4,\mathbb{R}))$$

while it acts directly on LM on the right according to

$$\underset{\sim}{E} \mapsto (E_\beta P^\beta{}_\alpha)\alpha = 1\ldots 4 \in LM.$$

2.7 Cross-sections

A cross-section of a bundle X with projection $\pi : X \to M$ is a continuous map $\sigma : M \to X$ such that $\pi\circ\sigma$ is the identity on M. We have already met classical fields as examples of cross-sections of bundles. A local cross-section is a map $\sigma : W \to X$ such that $\pi\circ\sigma$ is the identity on W.

Given a local cross-section $\sigma : W \to \pi^{-1}(W)$ of a principal fibre bundle X with group G, we can define a trivialisation θ of $\pi^{-1}(W)$ as follows. For any $q \in \pi^{-1}(x)$, with $x \in W$, since the action of G is simply transitive there is a unique $g_q \in G$ such that $q = \sigma(x) g_q$.

So define θ by $\theta(q) = (x, g_q) \in W \times G$.

This argument shows that in any bundle on which G acts simply transitively, cross-sections give trivialisations, and hence that any such bundle is a principal bundle.

2.8 Bundles with structure: sheaves

It may happen that each fibre of a bundle has some additional structure, such as being a vector space, a group, or an algebra. In this case we refer to the bundle as a vector bundle, group bundle, etc., provided that the additional structure fits in with the bundle-structure. Explicitly, we require that

(i) the standard fibre F has this structure;

(ii) the group acts on F as a group of automorphisms of the structure

(iii) the restriction of any coordinate map θ_i to a fibre defines an isomorphism between the structure on the fibre and the structure on F.

We regard two bundles with structure over M as equivalent if there is a map from one to the other mapping each fibre of one over x isomorphically onto the fibre of the other over x.

Now in physics the important object is not the fibre bundle X but the set of fields, $\mathscr{F}(X,M)$ consisting of all cross-sections $M \to X$. If X has an algebraic structure, then so has $\mathscr{F}(X,M)$. For example, if X is a vector bundle then $\mathscr{F}(X,M)$ is a vector-space, the vector operations being defined pointwise. In quantum mechanics the fields in $\mathscr{F}(X,M)$ are given a primary role, and one looks for representations of these fields, while the individual points in the bundle may be almost meaningless. In some cases one need not require the existence of global fields, but only local fields $\mathscr{F}(X,U) = \{\text{local cross-sections } U \to X\}$ for small enough U.

The question arises, can one work entirely with the objects $\mathscr{F}(X,U)$, for various U, as algebraic objects (vector spaces, groups etc, depending on the structure of X) without ever referring to the bundle X itself (since X may not mean anything physically)? The answer appears to be, yes. Suppose one is given a collection $\{\mathcal{F}(U) \mid U \subset M\}$, where U has to be open and each $\mathcal{F}(U)$ is, say, a vector space (or a group etc.). If one imposes appropriate axioms on how $\mathcal{F}(U_1)$ and $\mathcal{F}(U_2)$ have to be related for different U_1 and U_2 then one obtains a structure which obeys the same algebraic rules as an $\mathcal{F}(X,U)$, without referring to a particular X. The result is called a sheaf, or a presheaf. (The terminology differs: some authors reserve 'sheaf' to refer to a topological space that encodes the various $\mathcal{F}(U)$ for all U, while others use both words to refer to systems of $\mathcal{F}(U)$ satisfying slightly different axioms one from another).

The sheaf approach has more scope than the fibre bundle approach. For one thing, the spaces $\mathcal{F}(U)$ might constitute a special subset of the space of all $\mathcal{F}(X,U)$ (e.g. analytic cross-sections with respect to a complex structure); or it could be that there is no possible X which reconstructs a given set of $\mathcal{F}(U)$. So in the case of super-gravity it is particularly interesting to know that the two approaches are in fact equivalent, as shown in the notes of Batchelor and Rawnsley in this volume. Every sheaf of "superfields" is a sheaf of sections of a fibre bundle.

2.9 Associated bundles

Many physical fields take values in vector bundles rather than principal bundles; but for each such vector bundle one can find a principal bundle that bears the same relation to it as the frame bundle does to the tangent bundle. Geometrical structures become clearer in the principal bundle than in the vector bundle.

The easiest way to describe the relations between the two is to start with a principal bundle (E,π,M,G), and suppose given a representation $\rho : G \to \mathrm{Aut}(V)$ in some topological space V. That is, $\rho(g)$ is a homeomorphism of V onto itself and

$$\rho(g_1 g_2) = \rho(g_1) \circ \rho(g_2) \qquad (2)$$

We then define the bundle associated to E by ρ as follows. First define an equivalence relation ~ on ExV by

$$(q,v) \sim (qg^{-1}, \rho(g)v) \qquad \text{for any } g \in G.$$

(Equation (2) ensures transitivity). Then the space of the associated bundle is the set of equivalence classes:

$$E^{\rho} = ExV/\sim$$

and the projection $\pi^{\rho} : E^{\rho} \to M$ is defined by

$$\pi^{\rho}([q,v]) = \pi(q)$$

(writing $[q,v]$ for the equivalence class of $(q,v) \in ExV$ under ~).

If we are given a local section ρ of E, then it defines a trivialisation of E^{ρ} as well as of E (2.7), as follows. Take $q \in (\pi^{\rho})^{-1}(x)$, so that $q = [r,v']$ for $r \in \pi^{-1}(x), v' \in V$. Then there is a $g \in G$ such that $r = \sigma(x)g$, and so

$$q = [\sigma(x)g, v'] = [\sigma(x), v]$$

where $v = \rho(g^{-1})v'$. Define $\theta(q) = (x,v)$. It can be checked that the representation of q as $[\sigma(x),v]$ is unique, and that this is indeed a trivialisation map for E^{ρ}.

As an example, we can rederive the tangent bundle as associated to the frame bundles as follows. Take ρ to be the canonical representation of $GL(4,\mathbb{R})$ in $\mathbb{R}^4$, E = LM. Then an element of E^{ρ} is an equivalence class of pairs $(\underset{\sim}{E}, \xi)$ with $\underset{\sim}{E} \in LM$, $\xi \in \mathbb{R}^4$, with

$$(\underset{\sim}{E}, \xi) \sim (\underset{\sim}{E}A, A^{-1}\xi) \qquad (A \in GL(4,\mathbb{R})).$$

(writing $\underset{\sim}{E}A$ for $(\underset{\alpha}{E}A^{\alpha}{}_{\beta})\ \beta = 1,\ldots,4$). Clearly we can identify this class with the tangent vector

$\underset{\alpha}{E}\xi^{\alpha}$, since $(\underset{\sim}{E}A)_{\alpha}(A^{-1}\xi)^{\alpha} = (\underset{\alpha}{E}\xi^{\alpha})$, so that E^{ρ} is essentially the same as the tangent bundle. Indeed, the representation as the equivalence class $[\underset{\sim}{E}, \xi]$ is the "old fashioned" definition of a tangent vector by specifying its components ξ in every basis $\underset{\sim}{E}$, along with the transformation law that ξ changes to $\rho(A^{-1})\xi$ when $\underset{\sim}{E}$ changes to $\underset{\sim}{E}A$.

If V is a vector space and each $\rho(g)$ is a linear map, then E^{ρ} is a vector bundle; this is the usual situation.

Another case is where $V = G$ and ρ is the adjoint representation ad:

$$ad(g)h \equiv ghg^{-1}$$

Then E^{ad} is a bundle of groups, the group structure being defined by

$$[q,g][q,g'] = [q,gg'].$$

We shall call E^{ad} the gauge bundle of E.

2.10 Connections

Consider a principal bundle (E,π,M,G) with G a Lie group and E a differentiable manifold, such that the right action of G on E is smooth.

If $q \in \pi^{-1}(x) \equiv E_x$, say, then the tangent space T_qE to E at q will contain the subspace V_q of vectors that are tangent to E_x. This V_q is called the vertical subspace at q. A horizontal subspace is any subspace H_q of T_qE such that $T_qE = H_q \oplus V_q$. The action of $g \in G$ on E maps H_q into a subspace g_*H_q of $T_{qg}E$. A connection is then defined to be an assignment of an H_q at every $q \in E$ such that $g_*H_q = H_{qg}$ for all $g \in G, q \in E$. From the definition of a horizontal subspace, if a connection is provided, then every vector $X \in T_qE$ can be written uniquely as $X = X_v + X_h$ where $X_v \in V_q$ and $X_h \in H_q$.

To handle a connection one needs a way of expressing the map $q \mapsto H_q$; or, equivalently, the map $\mathcal{B}: X \mapsto X_v$ (since this determines X_h and hence H_q). The condition that $g_*H_q = H_{qg}$ is best analysed by looking at the components of X_v (and hence the components of $\mathcal{B}$) in a basis whose transformation properties under g_* are known: such a basis is provided by the Lie Algebra of G.

We recall a few basic ideas about Lie Algebras. If L_h : $g \mapsto hg$ denotes the map of $G \to G$ produced by left multiplication, then L_h induces a map L_{h*} on tangent vector fields on G. A vector field X is called left-invariant of $L_{h*}X = X$ for all $h \in G$. The set of all the left-invariant vector fields on G is called the

Lie algebra of G, written $\underline{g}$. Suppose that (as in the principal bundle) G acts on a manifold E on the right, so that there is a map $k: E{\times}G \ni (e,g) \mapsto eg \in E$. We also have the projection map $p: E{\times}G \ni (e,g) \mapsto g \in G$, and both these are onto, so that k_*^{-1} and p_*^{-1} exist as maps of vector fields on E (resp. G) to vector fields on ExG. Then for each vector field $X \in \underline{g}$ there is a unique induced vector field $\bar{X}$ on E such that $p_*^{-1}X = k_*^{-1}\bar{X}$. An alternative characterisation is that $\bar{X}$ is the tangent to the trajectories in E of the 1-parameter group of transformations $\{\phi_t\}_{t\in\mathbb{R}}$ given by $\phi_t(e) = e.\exp(tX) \quad (e \in E)$.

The vector fields $\bar{X}$ are not unchanged by the action of G (essentially because the Lie algebra consists of left-invariant fields while G acts on the right). In fact $g_*\bar{X} = \overline{ad_*(g^{-1})X}$ where $ad_*(h)$ is the mapping of vector fields induced by the transformation $g \mapsto hgh^{-1}$ on G.

So now express $\mathcal{H}$ in terms of the fields $\bar{X}$ by defining $\omega: TE \to \underline{g}$ (a $\underline{g}$-valued 1-form on E)

by $$\mathcal{H}(X) = \overline{\omega(X)(q)} \quad (X \in T_qE). \tag{2'}$$

Then the g_*-invariance of H_q is guaranteed if we require that

$$\omega(g_*X) = ad_*(g^{-1})\omega(X). \tag{3}$$

This means that it is only necessary to specify ω at one point of each fibre, after which equation (3) will specify it at every other point. Note that, if G is represented as a group of matrices, and if $\underline{g}$ is identified with the tangent space at the identity in G, whose elements are in turn also written as matrices (that is, as "infinitessimal group elements") then in this matrix representation (3) takes the form

$$\omega(g_*X) = g^{-1}\omega(x)g \tag{3'}$$

The choice of H_q allows one to pass from an element in one fibre of E to the "same" element in a nearby fibre by moving in the plane of H_q (to a first order approximation). This is the basis of the definition of covariant differentiation, for which we refer the reader to standard texts (1), and to §3.5 below.

The set of all $\underline{g}$-valued 1 forms on E defined over $\pi^{-1}(x)$ and satisfying (3) is a linear space, finite dimensional if M and G are finite dimensional. But ω must in addition satisfy the condition that if $X \in V_q$ then $\mathcal{P}(X) = X$, which is equivalent to

$$\omega(\bar{X}) = X \qquad \forall X \in \underline{g} \tag{4}$$

Any $\underline{g}$-valued 1-form satisfying (3) and (4) is called a connection form: any connection form uniquely specifies a connection. The imposition of (4) in addition to (3) restricts the space of possible connection forms on $\pi^{-1}(x)$ to an affine space W_x, with underlying vector space the space of all forms satisfying

$$\omega(\bar{X}) = 0 \qquad \forall X \in \underline{g}.$$

The set $\bigcup_x W_x$ forms an affine bundle (a fibre bundle where each fibre is an affine space: see § 2.8), whose cross-sections are all the connections on E. In physicists terminology, this is the bundle of gauge-fields for E.

In practice one manipulates connections by using a section σ of E (inducing a trivialisation as in § 2.7) and specifying ω at each $\sigma(x) \in E$. Since the action of ω on vertical vectors is given by (4) one needs only to specify ω on any subspace of vectors complementary to $V_{\sigma(x)}$, and one can take for this the vectors $\sigma_*(X)$ for $X \in T_xM$. If we choose once and for all a basis γ_A for $\underline{g}$ then we can fix ω by means of the so-called Cartan connection forms ω^A, 1-forms on M, by the equation

$$\omega^A(X)\gamma_A = \omega(\sigma_*(X)) \tag{5}$$

(see figure 2)

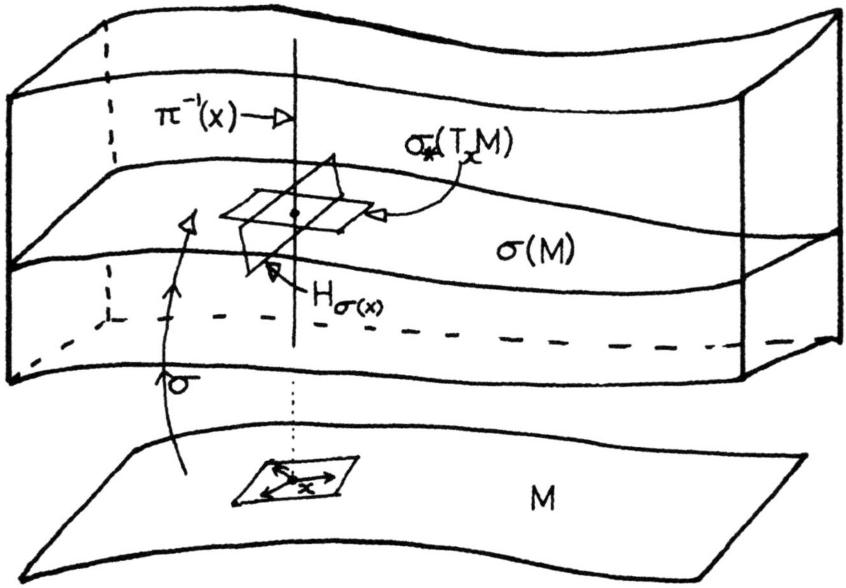

The trivialisation defined by σ allows us to refer to a point $q \in \pi^{-1}(x)$ by the pair (x,g) $(g \in G)$, where $q = \sigma(x)g$. If we represent G as a group of matrices and choose coordinates for x in M, then q has coordinates $(x^\mu, g^i{}_j)$. Consequently tangent-vectors to E can be written as $X = \eta^i{}_j \partial/\partial g^i{}_j + \xi^\mu \partial/\partial x^\mu$ (the first term in V_q, the second in $g_* \sigma_* (T_x M)$), while $\sigma(x)$ has coordinates (x^μ, δ^i_j). If the matrix representation of γ_A is $\gamma_A{}^i{}_j$ (regarding γ_A as an "infinitessimal group element", i.e. a tangent vector at the identity), then (5) and (2') give the following expression for the vertical component X_v of the vector $X = \eta^i{}_j \partial/\partial g^i{}_j + \xi^\mu \partial/\partial x^\mu$ in $T_{\sigma(x)}E$:

$$\varpi(X) = \varpi(X)^i{}_j \partial/\partial g^i{}_j$$

$$\varpi(X)^i{}_j = \xi^\mu \omega_\mu{}^A \gamma_A{}^i{}_j + \eta^i{}_j \qquad (6)$$

2.11 Examples

The situation for U(1)-electrodynamics is rather degenerate, since U(1) is not only Abelian but 1-dimensional. Because of the Abelian property, ad(h) is the identity and so (3) becomes

$$\omega(g_* X) = \omega(X).$$

Referring back to the gauge transformation in § 2.4, we note that the property $|\psi(x)| = 1$ is independent of the gauge, and so we can define E to be the set of all points in X having unit modulus in some (and hence any) gauge. U(1) obviously acts simply transitively on the elements of unit modulus and so, from the remark at the end of § 2.7, E is a principal bundle. Not surprisingly, the original X is its associated bundle.

With respect to a cross section σ of E, every element of E can be written as $e^{i\theta}\sigma(x)$ and so gives the coordinates (x,θ). The vector field $\partial/\partial\theta$ is a basis for the vertical vectors, on the Lie algebra index A takes the single value 1, $\omega^1_\mu = A_\mu$ (the vector potential) and (6) becomes

$$\varpi(\eta\partial/\partial\theta + \xi^\mu\partial/\partial x^\mu) = (\eta + A_\mu\xi^\mu)\partial/\partial\theta$$

In the case of the frame bundle LM the Lie algebra is isomorphic to the space of all 4 x 4 matrices and so the index A can range over all pairs (i,j) (i,j = 1,...,4), γ_A (A = (i,j)) being the matrix with components $(\gamma_i{}^j)^k{}_\ell = \delta^k_i\delta^j_\ell$. So (6) becomes

$$(\eta^i{}_j\partial/\partial g^i{}_j + \xi^\mu\partial/\partial x^\mu)^k{}_\ell = \eta^k{}_\ell + \omega_\mu{}^k{}_\ell\xi^\mu$$

(valid at coordinate $g^i{}_j = \delta^i_j$)

3 GAUGE GROUPS

Let (E,π,M,G) be a principal fibre bundle. The group of all homeomorphisms $\gamma: E \to E$ satisfying

$$\gamma(q)g = \gamma(qg) \qquad (\forall q\in E, g\in G) \tag{7}$$

is called the automorphism group of E, Aut(E). The subgroup of Aut(E) that preserves the fibres, so that $\pi(\gamma(q)) = \pi(q)$, will be called the gauge group of E, written $\mathcal{G}(E)$ or simply $\mathcal{G}$.

The reason for the name is expressed in the following result, which means that gauge transformations are the "active" versions of coordinate transformations.

3.1 Proposition: gauge transformations

Let $\sigma : U \to E$ be a local section and let the corresponding trivialisation be $q \mapsto (x, g_\sigma(q))$ where $q = \sigma(x)g_\sigma$. Take any $\gamma \in \mathscr{G}(E)$. Then there is another section σ' and associated $g_{\sigma'}$ such that

$$g_\sigma(\gamma(q)) = g_{\sigma'}(q).$$

In other words, the change in coordinates produced by the active transformation γ is the same as that produced by taking a different trivialisation, i.e. a different gauge.

The proof is trivial: there is a unique $k_\sigma(x) \in G$ such that $\gamma(\sigma(x)) = \sigma(x)k_\sigma(x)$. Define $\sigma'(x) = \sigma(x)k_\sigma(x)^{-1}$ so that $g_{\sigma'}(q) = k_\sigma(x)\ g_\sigma(q)$. Then

$$\gamma(q) = \sigma(x)g_\sigma(\gamma(q)) \qquad (*)$$

$$= \gamma(\sigma(x)g_\sigma(q)) = \gamma(\sigma(x))g_\sigma(q)$$

(from the definition of Aut(E))

$$= \sigma(x)k_\sigma(x)g_\sigma(q) = \sigma(x)g_{\sigma'}(q). \qquad (**)$$

whence the result on equating the r.h.s of (*) and (**).

The gauge group $\mathscr{G}$ consists precisely of all sections of the gauge bundle E^{ad} (section 2.9), in the following way. Let $\theta : M \to E^{ad}$ be such a section. Suppose $\pi(q) = x$ and write $\theta(x) = [q,h]$ ($q \in E, h \in G$, equivalence class under the adjoint representation). Then define any automorphism γ_θ on E by $\gamma_\theta(q) = qh$. To verify (7), showing that γ_θ is in the gauge group, we calculate that $\theta(x) = [qg, ad(g^{-1})h] = [qg, g^{-1}hg]$ and so

$$\gamma_\theta(qg) = qgg^{-1}hg \;=\; qhg \;=\; \gamma_\theta(q)g.$$

It can be shown, conversely, that every element of $\mathscr{G}(E)$ has the form of γ_θ for some section θ of E^{ad}.

3.2 Gauge action on associated bundles

If E^τ is associated to a principal bundle E by a representation τ of E, then it is easily verified that an action of each element γ of $\mathscr{G}(E)$ is defined by

$$\gamma : [q,v] \mapsto [\gamma(q),v].$$

If we use the notation of the previous section, we have

$$\gamma : [\sigma(x),v] \mapsto [\sigma(x)k_\sigma(x),v] \quad = \quad [\sigma(x),\tau(k_\sigma(x))v]$$

So the effect of γ is to act on the coordinate with respect to σ with $k_\sigma(x)$. This is consistent with the result of the previous section, that the effect of γ on E itself is given by

$$g_\sigma(\gamma(q)) \quad = \quad k_\sigma g_\sigma(q),$$

if we note that, if τ is left multiplication by G on G itself, then $E^\tau \approx E$.

It is easy to verify that if $\gamma = \gamma_\sigma$ for a section θ of the adjoint bundle E^{ad} with its group action defined in 2.9, then γ acts on the left, in the sense that

$$\gamma_\theta \gamma_\phi [q,v] = \gamma_{\theta\phi}[q,v].$$

3.3 Quasi-gauge groups

We saw in 3.1 that an element of the gauge group is a cross-section of a certain bundle of groups, which (cf. 3.2) can act on any associated bundle, in particular, on associated vector bundles. Suppose we generalise this to any bundle of groups (K,p,M,H,G') and any vector bundle (X,π,M,V,G), with an action

$$\gamma_x : p^{-1}(x) \times \pi^{-1}(x) \ni (h,q) \mapsto hq$$

linear on each fibre. Thus X can be thought of as the bundle-space of a set of vector fields, and K as a bundle of local symmetry-groups - one of the most basic situations in mathematical physics. We shall call the set of cross-sections of K a quasi-gauge group.

The natural question is, is every quasi-gauge group a gauge group? More precisely, are there always equivalences $\chi : E^{ad} \to K$, $\rho : E^\tau \to X$ (where E is a principal fibre bundle and τ is a representation of its group) such that χ and ρ are consistent with the group actions of E^{ad} on E^τ and of K on X?

For this to be true we require that if we have a

trivialisation $\theta : p^{-1}(w) \to W\times H$ of Γ over a neighbourhood W in M, and if τ_x is the action of H on $\pi^{-1}(x)$ defined by

$$\tau_x(h)(v) = \theta^{-1}(x,h)(v) \qquad (x \in M, h \in M, v \in X_x)$$

then the two representations τ_x and τ_y must be equivalent for any $x,y \in W$. With this condition one can show that locally (over any contractible neighbourhood) our question can be answered in the affirmative: there exists a principal fibre bundle locally whose gauge group is equivalent to K.

But globally this is not true, as is shown by a physically interesting counter-example.

Suppose space-time M is not simply connected and is not space and time orientable (i.e. it is possible to describe a path in space-time, returning to one's starting point having experienced a spatial and a temporal reflection - combined with a charge conjugation, if CPT invariance is to be preserved). Then suppose we try to construct a field theory in which the fields are cross sections of the tangent bundle of space-time, X = TM, while the theory is invariant only under the group K of local Lorentz transformations in TM preserving the space and time orientations. In other words, we try to implement a theory whose symmetry is smaller than the group of transformations naturally arising from the space-time. There is nothing to stop us defining a pseudo-gauge-group K to do this, but it is easy to show that it can never be a gauge group in the sense in which we have defined it.

3.4 Gauge algebras

In practice it is normal to work with "infinitessimal gauge transformations", or elements of the gauge algebra. To define this, take K to be either E^{ad} for a principal fibre bundle, or the bundle of pseudo-gauge group as in the previous section, with each fibre of K acting linearly on the corresponding fibre of a vector bundle X. If k_x denotes the Lie algebra of the Lie group $K_x = p^{-1}(x)$, then $\bigcup_x k_x$ is a bundle of Lie algebras.

Now in section 2.10 we showed how, given a Lie group acting on a manifold, there was induced a Lie algebra of vector fields on the manifold (the trajectories of the action). Let $\bar{k}_x$ be the algebra of fields on $X_x = \pi^{-1}(x)$ induced by the action of $K_x = p^{-1}(x)$. Since K_x acts on the left, $\bar{k}_x$ is naturally isomorphic the algebra of right-invariant vector fields on K_x.

Let ξ be a cross-section of $\bar{k} = \bigcup_x \bar{k}_x$. Then each $\xi(x)$ defines an element of $\bar{k}_x$, i.e. a vector field on X_x, and so ξ as a whole defines a vector field V_ξ on X, with the fibres of X being integral manifolds of V_ξ. From the definition, the commutator of two fields V_ξ and V_η is the field $V_{[\xi,\eta]}$, where $[\xi,\eta]$ is evaluated in the Lie-algebra structure of the $\bar{k}_x$. Thus we define the set of vector fields $\Xi = \{V_\xi : \xi$ is a cross-section of $\underline{k}\}$ to be the gauge algebra of K (the algebra of infinitessimal gauge (or pseudo gauge) transformations.

3.5 Gauge-invariance

If X is a fibre bundle over M we denote by $\mathscr{F}(X,M)$ the set of cross-sections (fields), as in §2.7. If a group $\mathscr{G}$ acts on X in such a way that, for each $g \in \mathscr{G}$, $\phi \in \mathscr{F}(X,M)$ and $x \in M$ we have that there exists a single point $p(g,\phi,x)$ with

$$g(\phi(M)) \cap \pi^{-1}(x) = \{p(g,\phi,x)\}$$

then we can regard g as mapping every section ϕ into a section $g(\phi)$ defined by

$$g(\phi)(x) = p(g,\phi,x).$$

In most field theories we are concerned with functions defined on the fields, e.g. $S:\mathscr{F}(X,M) \to \mathbb{R}$, such as the action function. We say that such an S is invariant under $\mathscr{G}$ if

$$S(g(\phi)) = S(\phi) \qquad (\forall \phi \in \mathscr{F}(X,M), g \in \mathscr{G}).$$

Where $\mathscr{G}$ is the gauge group, a theory satisfying this condition is said to be gauge-invariant.

Suppose now that $\{\theta_\tau\}_{\tau \in \mathbb{R}}$ is a one-parameter subgroup of a group $\mathscr{G}$ acting on a vector bundle X and let V be the vector field tangent to the trajectories of the action of $\{\theta_\tau\}$; i.e.

$$(Vf)(x) = \frac{d}{d\tau} f(\theta_\tau(x))\Big|_{\tau = 0}$$

for all functions $f : X \to \mathbb{R}$. Then we can define a sort of Lie derivative of ϕ by the formula

$$(\pounds_V\phi)(x) = -\frac{d}{d\tau}\theta_\tau(\phi)(x)\Big|_{\tau = 0} \in X_x$$

(where the right hand side can be shown to depend only on V, ϕ and x). Its value represents the infinitessimal change of ϕ under the subgroup θ_τ.

Since in this case $\mathscr{F}(X,M)$ is a vector space, this change in ϕ induces an infinitessimal change in S given by

$$\frac{d}{d\tau} S(\theta_\tau(\phi))\Big|_{\tau = 0} = S'(\phi)(\pounds_V\phi)$$

where $S'(\phi)$ is the Frechet derivative of S at ϕ, with respect to a suitable topology on $\mathscr{F}(X,M)$. If S is invariant under $\mathscr{G}$, this gives

$$S'(\phi)(\pounds_V\phi) = 0 \tag{8}$$

for all $\phi \in \mathscr{F}(X,M)$ and all V tangent to one-parameter subgroups of $\mathscr{G}$. When $\mathscr{G}$ consists of sections of a (pseudo-) gauge bundle with gauge algebra Ξ (as in the previous section) then the equation holds for all $V \in \Xi$, i.e. for all infinitessimal gauge transformations.

The general situation is where θ_τ is a subgroup of the automorphism group (see 3.0), not restricted to being a gauge transformation in our sense.

(Warning In supergravity theories all the transformations leaving S invariant are referred to as "gauge transformations", whether they keep the fibres fixed - and so are gauge transformations in our sense - or whether they are members of the automorphism group moving the fibres.)

We can describe the situation also in terms of the vector fields V on X, without reference to a group of transformations that induces them. This is particularly valuable if one only has local infinitessimal symmetry transformations without knowing that they exponentiate up into a full group (a process that may be hindered by topological complications in spacetime and algebraic problems in superspace). To formulate invariance in this way, first take a trivialisation θ of part of the bundle,

$$\theta : \pi^{-1}(W) \rightarrow W \times F$$

Write $\theta_x V = (V_1, V_2)$ where V_1 and V_2 are vector fields on W and F respectively, $\bar{\phi} = p_2 \circ \theta \circ \phi$ where p_2 is projection on F. So we now have a field $\bar{\phi}$ taking values in F and a "gauge" transformation composed of a space-time displacement V_1 and a displacement in F of V_2. Then agreement with the previous definition of $\mathcal{E}_V$ is obtained if we define $\mathcal{E}_V$ by

$$\overline{\mathcal{E}_V \phi} = V_2 - V_1 \bar{\phi}$$

Suppose now that Σ is a set of vector fields such that (8) holds for any $V \in \Sigma$, with this definition of $\mathcal{E}$. Then it can be calculated directly that, if V and W are in Σ, then (8) holds for their commutator [W,V] (provided there is sufficient differentiability for everything to be defined and that S is twice differentiable). Thus we may suppose that the set Σ is completed to a Lie algebra. In this case S is said to be invariant under Σ.

The existence of an algebra under which S is invariant gives rise to conserved currents (Noether's theorem), which has been described by Fischer (3) in the geometrical context of the gauge- and automorphism-groups.

We postpone examples until after the section on space-time.

3.6 Gauge theory

In the previous section a connection was used as a background reference-geometry. In a gauge theory, the connection is itself a dynamical field and the bundle of all the fields has the form

$$X = \bigcup_x X_x, \quad X_x = Y_x \oplus W_x$$

where $Y = \bigcup_x V_x$ is a vector bundle and $W = \bigcup_x W_x$ is the affine bundle of all connections on a principal bundle E to which Y is associated. In this case we write $X = Y \oplus W$. Then it turns out that any subgroup of the automorphism group of E automatically extends uniquely to a group of automorphisms of X. If we have a 1-parameter family θ_τ acting on E, and hence on Y, and if V is tangent to its trajectories, then when we extend the θ_τ to act on W, the bundle of connections, then the $\pounds$-derivative becomes simply the usual Lie derivative:

$$\pounds_V \omega = \mathcal{L}_V \omega$$

where on the left ω is regarded as a cross section of W, on the right as a 1-form on E.

If the action in a gauge theory involves only the Y-fields and their covariant derivatives with respect to the connection (W-fields), then the action will automatically be gauge-invariant; indeed, invariant under Aut(E). This gives the fundamental method of constructing invariant actions in gauge theories. We may introduce any subsidiary structure (metric, 4-forms on M) provided it is invariant under a subgroup $\mathcal{G}$ of Aut(E), in which case the resulting action will be $\mathcal{G}$-invariant.

4. SPACE-TIME

4.1 Spinors

The traditional approach to spinors in general relativity (4) proceeds in three stages. First, the metric on space-time is specified, allowing one to construct the bundle of all frames that are pseudo-orthonormal with respect to the metric. Next a spin structure is introduced, if possible, which allows one to construct a two-fold covering of this bundle with $SL(2,\mathbb{C})$ as structure group. Finally, various associated bundles are constructed in which the spinor fields exist as cross sections. In this approach, therefore, it is not possible to start off by regarding the metric and the spinor fields as dynamical entities on the same footing, because the metric has to be fixed before it is even possible to talk about spinor fields. So I shall describe here an alternative approach which underlies many supergravity theories, in particular that described by Isenberg in these proceedings.

We begin with a principal bundle E over a 4- manifold M having $SL(2,\mathbb{C})$ as structure group. At this stage M has no metric so that M cannot properly be called space-time and E cannot properly be called a spin bundle. E^{o} will be the vector bundle associated to E by the standard representation of $SL(2,\mathbb{C})$ in $\mathbb{C}^2$, while $E^{\cdot}$ is the vector bundle associated by the representation

$$SL(2,\mathbb{C}) \ni A \mapsto A^{*}:\mathbb{C}^2 \to \mathbb{C}^2,$$

where * denotes complex conjugation. There are natural maps $\bar{}$: $E^{o} \to E^{\cdot}$ and $E^{\cdot} \to E^{o}$ given by $\eta = [q,v] \to \bar{\eta} = [q,v^{*}]$.

Note that we are free to regard $\mathbb{C}^2$ throughout this development either as a two-dimensional complex vector space, resulting in E^{o} and $E^{\cdot}$ being complex vector bundles, or as a four-dimensional real vector space, in which case E^{o} and $E^{\cdot}$ will be real vector bundles. In the latter case elements of E^{o} and $E^{\cdot}$ will become 'Majorana spinors', once a metric has been determined.

With respect to a section $\sigma : M \to E$, coordinates in E^{o} and $E^{\cdot}$ behave as follows. In the complex case the coordinates of $\eta = [\sigma(x),v] \in E^{o}$ are just (v^1,v^2), while those of $\bar{\eta}$ are (v^{1*},v^{2*}), so that

$$\bar{\eta}^{\dot{A}} = \eta^{A*} \quad (A = 1,2) \qquad (\eta^{A} \text{ complex})$$

adopting the usual convention of indexing components of elements of $E^{\cdot}$ by $(\dot{1},\dot{2})$. In the real case we use the basis $\{(1,0),(0,1),(i,0)(0,i)\}$ of $\mathbb{C}^2$ as a real vector space to obtain

$$\bar{\eta}^{\dot{\alpha}} = J^{\dot{\alpha}}{}_{\beta}\eta^{\alpha} \qquad (\alpha = 1,\ldots,4) \; (\eta^{\alpha} \text{ real})$$

where $[J^{\dot{\alpha}}{}_{\beta}] = \mathrm{diag}(1,1,-1,-1)$. There are various alternative descriptions of this case corresponding to different choices of a basis for $\mathbb{C}^2$ as a real vector space, among them the "Majorana spinors" used for supergravity.

4.2 Soldering forms

Denote the projection $E \to M$ by π. Then an $\mathbb{R}^4$-valued 1-form e on E will be called a soldering form if

(i) $e(X) = 0$ for all $X \in TE$ with $\pi_* X = 0$

(ii) $e_q : T_q E \to \mathbb{R}^4$ is surjective for each $q \in E$

(iii) For each $A \in SL(2,\mathbb{C})$, $R_{A*}\overset{b}{e} = (P(A))^{b}{}_{a}\overset{a}{e}$

where P(A) is the Lorentz transformation corresponding to A, $R_A : q \mapsto qA$ is the right action of $SL(2,\mathbb{C})$ on E, and the $\overset{a}{e}$ are the four components of e (each a real 1-form).

Given e, we can associate a frame $E_e(q)$ on the base M with each $q \in E$, as follows. Given $X \in T_x M$ with $x = \pi(q)$, set

$$X^a(q) = \overset{a}{e}(q)(\bar{X})$$

where $\pi_* \bar{X} = X$. From (i), X^a is independent of the choice of $\bar{X} \in T_q E$. Then (ii) implies that the map $X \mapsto X^a$ is surjective, so that the frame $E_e(q) = (E_{0e}(q),\ldots,E_{3e}(q))$ can be defined by the requirement that $E_{ae}(q)X^a(q) = X$. Condition (iii) implies that under a change from q to qA the components of X change to $P(A)^b{}_a X^a$, and so the frame transforms contragrediently according to

$$(E_a)_{a=0,\ldots,3} \mapsto (E_b P(A)^b{}_a)_{a=0,\ldots,3}$$

The frames $\{E_e(q) : q \in E\} = L_e M$ form a Lorentz bundle which is a sub-bundle of the frame bundle over M. If we were to follow the procedure outlined at the start of the previous section, constructing a bundle of spinors associated to L_e we should recover the original E^o.

The Lorentz bundle $L_e M$ fixes a metric g_e on TM, where as usual we set $g_e(X,X) = +(X^0)^2 - (X^1)^2 - (X^2)^2 - (X^3)$ for components with respect to any $q \in p_E^{-1}(x)$ $(\pi X = x$, where $\pi : TM \to M)$.

The map $q \mapsto E_e(q)$ defines a two-fold covering of the frame bundle by the original bundle E, and the image of e under this map is precisely the "soldering form" of ordinary differential geometry, so called because it "sticks down" the fibres onto the manifold M by relating them to the tangent vectors on M.

4.3 Achtbeine

The right-covariance of the soldering form, expressed by (iii) of the preceding section, implies that it is only necessary to specify e at the points of a cross-section of E, just as happened for the specification of the connection. If σ is such a cross section, then the frame $E_e(q)$ associated with $\sigma(x) = q$ is given by $E_e(\sigma(x)) = \sigma^*(e(\sigma(x))) \equiv K_e^\sigma(x)$, say.

Turn now to the associated vector bundle E^0, using the action of SL(2,C) on $\mathbb{C}^2_{\mathbb{R}} \simeq \mathbb{R}^4$ as a real vector space, so that E^0 is a 4-dimensional real vector bundle. If $\pi^0 : E^0 \to M$ is the projection, it induces the $\mathbb{R}^4$ - valued 1-form $\pi^{0*}K_e^\sigma = F_e^\sigma$, say, on E^0, with components $\overset{0\sigma}{F}_e, \ldots, \overset{3\sigma}{F}_e$.

The point $\sigma(x)$ can be regarded as equivalent to the frame $Z^\sigma(x) = (\underset{1}{Z}(x), \ldots, \underset{4}{Z}(x))$ on the fibre $E_x^0 = \pi^{0-1}(x)$ defined by

$$\underset{i}{Z}(x) = [\sigma(x), C_i]$$

where $C_1 = (1,0)$, $C_2 = (0,1)$, $C_3 = (i,0)$, $C_4 = (0,i)$ are a real basis for $\mathbb{C}^2_{\mathbb{R}}$ and [,] denotes the equivalence class in $E \times \mathbb{C}^2$ under the $SL(2,\mathbb{C})$ action used in defining E^0. Since E_x^0 is a vector space, for each point $v \in E_x^0$ each vector $\underset{i}{Z}^\sigma(x)$ can be canonically identified with a vector

$$\dot{Z}^\sigma_{iv} \in T_v(E_x^0).$$

An achtbein is a set of eight 1-forms $(\varepsilon_1, \ldots, \varepsilon_8)$ on E^0 such that

$$\text{(i)} \qquad \varepsilon_i(v) = \overset{i-1}{F}{}^{\sigma}_{e}(v) \qquad i = 1,\ldots,4$$

$$\text{(ii)} \qquad \varepsilon_m \cdot Z^{\sigma}_{nv} = \delta_{m-4,n} \qquad M = 5,\ldots 8; \quad n = 1,\ldots,4$$

Where parentheses in (i) denote evaluation at v and the dot in (ii) denotes the vector space/dual space pairing for $T_v(E^o)$.

Note that these conditions do not specify the achtbein uniquely: to do this we should also have to fix the space of vectors $V \in T_v(E^o)$ for which

$$\varepsilon_m(V) = 0 \qquad (m = 5,\ldots,8).$$

It might be natural to take for this space the horizontal subspace corresponding to the torsion-free (Levi-Civita) connections induced by e.

An analogue of this achtbein is used in the superspace formulation of super-gravity, used by Professor Wess in his lectures in this volume. Here E^o becomes a super-manifold: one still has the structure of a fibre-bundle over space-time M, but the fibres are Grassmanian spaces. And the achtbein is not restricted by the conditions under which we derived it, but is restricted by the requirement that it be differentiable with respect to the Grassmanian super-manifold structure, which forces at most an affine dependence on the coordinates in the fibres. Further restrictions are imposed by placing requirements on the metric defined on E^o.

4.4. Example: Lie derivatives

Before discussing a situation more closely related to supergravity, let us consider symmetries on the bundle LM of all frames for the tangent bundle of a manifold M, a $GL(n,\mathbb{R})$- bundle. If we are given a section of LM (i.e. a field of frames) then every motion of M gives rise, by Lie dragging, to a motion of the field of frames.

Expressed in local coordinates ξ^{μ}_{a} on the standard fibre $GL(n,\mathbb{R})$ and x^{μ} on M, if we are given a vector field with components X^μ on M, the change in a field of frames $x \mapsto (e^\mu_a(x))_{a,\mu}$ produced by the diffeomorphism $\phi_{\delta t}$ obtained by moving a parameter δt along the integral curves of X is

$$\delta e^\mu_a = -\delta t(\mathcal{L}_X e)_a + O(\delta t^2)$$

i.e.

$$\frac{de^\mu_a}{dt} = -X^\mu \frac{\partial e^\mu_a}{\partial x^\nu} + e^\nu_a \frac{\partial X^\mu}{\partial x^\nu}$$

$$= (\mathfrak{L}_{\tilde{X}} e)^\mu_a$$

in the notation of section 3.4, where $\tilde{X}$ is the vector field on LM given in components as

$$\tilde{X} = \xi^\nu_a \frac{\partial X^\mu}{\partial x^\nu} \frac{\partial}{\partial \xi^\mu_a} + X^\nu \frac{\partial}{\partial x^\nu} .$$

Thus a vector field ("infinitessimal diffeomorphism") on M lifts to a vector field $\tilde{X}$ on LM which, when applied to fields as in 3.4 has the effect of Lie-dragging. The requirement that a theory be invariant under diffeomorphisms then translates into the requirement that it is invariant under the algebra of all fields on LM of the form $\tilde{X}$.

4.5 Supersymmetries

We briefly mention this as an illustration of a gauge algebra element and a link with the theory of supergravity. First we need to define the bundle where the Rarita-Schwinger fields live.

Let $\tilde{P}$ be the representation of SL(2,$\mathbb{C}$) in $\mathbb{C}^2_{\mathbb{R}} \otimes \mathbb{R}^4$ given by

$$\tilde{P}(A)(\eta \otimes v) = (A\eta) \otimes v\, P(A)$$

where P is the Lorentz representation of 4.2.

The Dirac map

$$\gamma : \mathbb{C}^2_{\mathbb{R}} \otimes \mathbb{R}^4 \to \mathbb{C}^{2\,*}_{\mathbb{R}}$$

$$\eta^A \xi_\mu \mapsto \gamma^\mu_{BA} \eta^A \xi_\mu$$

(where * denotes the algebraic real dual) is invariant in the sense that

$$\gamma(\tilde{P}(A)\psi) = \gamma(\psi)A^{-1}$$

and so $\tilde{Q} = \ker\gamma$ is an invariant subspace of $\mathbb{C}^2_{\mathbb{R}} \otimes \mathbb{R}^4$. In fact the representation of $SL(2,\mathbb{C})$ defined by restricting $\tilde{P}$ to $\tilde{Q}$ is irreducible, of spin $^3/_2$.

Let Q be the vector bundle associated to E by this representation: then its cross-sections are the Rarita-Schwinger fields.

Because of the covering of the frame bundle E_e by E, Q can be identified with a sub-bundle of $E^o \otimes T^*M$. More explicitly, we can define a map

$$\lambda^e : [q,\mu\otimes\xi] \mapsto [q,\eta] \otimes [E_e(q),\xi]$$

which we can then restrict to

$$Q \to E^o \otimes T^*M.$$

Just as the set of connections forms an affine bundle (section 2.10), so the set of soldering forms forms a fibre bundle H over M, though without an affine structure. The fibre-wise Cartesian product $H \oplus Q$ is the commuting version of the space of fields for N = 1 supergravity.

Supersymmetries (infinitessimal) cannot be expressed as vector fields on $H \oplus Q$ (even ignoring the question of anticommutativity) in the way that infinitessimal diffeomorphisms were so expressed in 4.4, because their formulation involves the connection, which depends on the derivatives of the soldering field e in a way that cannot be expressed as an $\pounds\phi$. So one must include the bundle of connection forms explicitly in order to formulate them. But it turns out that, if one imposes the condition that the fields in Q anticommute, then the commutator of two supersymmetries can be written as an $\pounds_Z\phi$ in $H \oplus Q$; indeed, the form of Z is very close to that which appears in the expression for the Lie derivative vector-field $\tilde{X}$ of the previous section. It is in this sense that the commutator of two supersymmetries is like an infinitessimal diffeomorphism. But the expression of this requires us to leave the confines of ordinary (commutative) differential geometry to which these lectures have been confined.

References

1. Kobayashi, S. and Nomizu, K. "Foundations of differential geometry", Interscience, New York, 1963
2. Steenrod, N. "The topology of fibre bundles", Princeton University Press, Princeton, 1951
3. Fischer, A. "Isotropy groups as universal symmetry groups in general relativity" to appear in GRG-Journal 1984
4. Lichnerowicz, A. in "Battelle Rencontres: 1967 Lectures in Mathematics and Physics", Eds. Dewitt, C.M. and Wheeler, J.A., Benjamin, New York, 1968, pp. 107-116.

PATH INTEGRATION ON MANIFOLDS

K.D. Elworthy

Mathematics Institute, University of Warwick, Coventry.

These notes form a fairly standard introduction to Wiener integration on $\mathbb{R}^n$ and on Riemannian manifolds. Feynman path integrals for non-relativistic quantum mechanics are also considered and compared to Wiener integrals. The basic approach is via cylinder set measures, Gaussian measures, and abstract Wiener spaces.

H.-J. Seifert et al. (eds.), Mathematical Aspects of Superspace, 47–89.

1. INTRODUCTION

Superspace does not figure in these notes outside of the references and this introduction. At the time of writing them I do not know of a mathematical theory of path integration on superspace or supermanifolds although recent developments suggest that it could be very profitable to mathematics to have such a theory. One of these developments is the approach to the Atiyah-Singer index theorem for the Dirac operator via supersymmetric quantum mechanics and the Witten index by Alvarez-Gaumé and D. Friedan & P. Windey. Another example is the Parisi-Sourlas result on dimensional reduction and supersymmetry for a system in a random magnetic field, with a rigorous mathematical version promised in (26). Some simple examples of path integration in superspace are worked out by B.S. DeWitt in his forthcoming monograph (6) which gives a clear introduction to the subject. Reference (4) on supersymmetric quantum mechanics may also be found useful.

What is in these notes is a mathematical description of fairly standard material on path integration on $\mathbb{R}^n$ and on finite dimensional Riemannian manifolds M. The main concern is with Wiener integration: 'imaginary time' or 'Euclidean theory' from the point of view of quantum mechanics. However Feynman path integrals are also discussed at some length and their formal similarity with Wiener integrals is brought out both for paths on $\mathbb{R}^n$ and for paths on M . In fact a major aim of the presentation here has been to give a general setting which is likely to be of use in a variety of situations, in particular for path integration on superspace and supermanifolds.

The setting is also relevant for field theory, particularly Euclidean field theory although these aspects are also not mentioned. Relevant references to this can be found in (8). Path integration on infinite dimensional manifolds is applied to Euclidean gauge field theory in (3) and (21) and discussed in general in (13). Other topics not discussed include phase space path integrals (31), or (10) for a heuristic treatment, and the Poisson process approach to path integration in phase space described by Combe et al. in (8).

For more details, background, and the bibliographical facts see the "Stepping Stone" in (8), although manifolds do not figure in it, or the lecture notes (7). Both of these were designed for non-mathematicians: the full details etc., of the manifold theory appear in (15). A quick summary with examples of applications of path integration to differential geometry can be found in (14).

The notes are arranged as follows: §2 discusses generalities about measures on infinite dimensional linear spaces, especially Gaussian measures. It defines Wiener measure on the space of paths on $\mathbb{R}^n$ and describes the Feynman-Kac formula for solutions of the diffusion equation on $\mathbb{R}^n$, expressing the solution also a limit of Feynman type sums via a time slicing procedure. Feynman path integrals are described in §3 together with their relationship with the Schrödinger equation in non-relativistic quantum mechanics. In §4 we consider Wiener and, briefly, Feynman integration for paths on a Riemannian manifold M , giving the construction of 'Wiener measure' for paths on M via the 'stochastic development' and stating the corresponding Feynman-Kac formula. The relationship of the time sliced version of this formula with the Pauli-Van-Vleck-DeWitt approach to Feynman path integration on M is also described. In §5, the last section, we consider path integration for differential forms on M , and gauge invariant path integrals.

2. GAUSSIAN MEASURES, CYLINDER SET MEASURES, AND THE FEYNMAN-KAC FORMULA

2.1 Basic difficulties; terminology

There are two major setbacks to the rigorous mathematization of standard practices by quantum physicists:

(i) $(2\pi i)^{-\frac{1}{2}n} \exp \frac{1}{2} i|x|^2 \, dx$ does not determine a complex measure on $\mathbb{R}^n$. In particular the integrals over $\mathbb{R}^n$ of functions like $|x|^2 \exp(\frac{1}{2} i|x|^2)$ have no meaning (until some regularization procedure is laid down);

(ii) there are no measures on infinite dimensional spaces

which behave like Lebesgue measure on $\mathbb{R}^n$.

Some explanation of the terminology is needed: consider a set X and a family $\mathcal{A}$ of subsets of X which forms a *σ-algebra* (i.e. countable unions, countable intersections, and complements, of sets in $\mathcal{A}$ lie in $\mathcal{A}$, and $X \in \mathcal{A}$). Then $(X,\mathcal{A})$ is a *measurable space* and a *measure* μ on $(X,\mathcal{A})$ is a map

$$\mu:\mathcal{A} \to [0,\infty) \cup \{+\infty\}$$

$$A \mapsto \mu(A)$$

such that if A_i , $i = 1$ to ∞ , are disjoint and in $\mathcal{A}$ then

$$\mu\left(\bigcup_{i=1}^{\infty} A_i\right) = \sum_{i=1}^{\infty} \mu(A_i) .$$

It is a *finite measure* if $\mu(X) < \infty$. When X is a topological space we shall always take $\mathcal{A}$ to be the σ-algebra of *Borel sets:* the smallest σ-algebra which contains all the open subsets of X .

A map $\theta:X \to Y$ where $(X,\mathcal{A})$ and $(Y,\mathcal{B})$ are measurable spaces is *measurable* if $\theta^{-1}(B) \in \mathcal{A}$ whenever $B \in \mathcal{B}$. Given such a map θ and a measure μ on $(X,\mathcal{A})$ there is a measure $\theta(\mu)$, or $\mu\circ\theta^{-1}$, induced on $(Y,\mathcal{B})$ by

$$\theta(\mu)(B) = \mu(\theta^{-1}B) \qquad B \in \mathcal{B} .$$

This has the important property that for $f:Y \to \mathbb{C}$

$$\int_X f(\theta(x))d\mu(x) = \int_Y f(y)d(\theta(\mu))(y) \qquad (1)$$

whenever the integrals exist.

For a measure space $(X,\mathcal{A})$ a map $\mu:\mathcal{A} \to \mathbb{C}$ is a *complex measure* (of absolute bounded variation) if there is a measurable $\theta:X \to \mathbb{R}$ and a finite measure $|\mu|$ on $(X,\mathcal{A})$ with

$$d\mu = e^{i\theta(\cdot)}d|\mu|$$

$$\mu(A) = \int_A e^{i\theta(x)}\, d|\mu|(x) \qquad A \in \mathcal{A} .$$

For example if X is the positive integers the assignment of weight $(-1)^n 1/n$ to the singleton set $\{n\}$ of X will not determine a complex measure. This agrees with intuition because the 'total mass' of X would depend on the order in which the terms are weighed (since the series $-1 + \frac{1}{2} - \frac{1}{3} + \frac{1}{4} \cdots$ can be rearranged to sum to any preassigned real number).

Assertion (i) above is therefore true by definition of a complex measure. For assertion (ii) we take the basic properties of Lebesgue measure to be those of translation invariance and *local finiteness* (each point is contained in some open set with finite measure). In fact there is no locally finite measure μ on a separable Banach space E which is quasi-invariant under translation by all elements of E (μ is *quasi-invariant* under translation by $x \in E$ if the measure $\mu(x+A)$, of the translate of a Borel set A of E, vanishes if and only if $\mu(A) = 0$). For a simple proof of this in the context of measures on topological groups see (23).

One way round the difficulties stemming from (i) is to 'go Euclidean', or 'change to imaginary time', by considering $(2\pi iz)^{-n/2}\exp(\frac{i|x|^2}{2z})dx$ for $z = -i$ rather than for $z = 1$. This leads to the study of Gaussian measures, and it so happens that these also seem to be the best replacements for the non-existent Lebesgue measures on infinite dimensional spaces.

2.2 Gaussian Measures

Suppose V is a real vector space with $\dim V = n < \infty$. A *non-degenerate Gaussian measure* γ on V is a measure on V such that for all Borel sets A of V

$$\gamma(A) = \int_A (2\pi)^{-n/2} \exp(-\tfrac{1}{2}|x|^2)dx$$

for some inner products $< , >_\gamma$ on V with corresponding norm $|\ |_\gamma$ and resulting Lebesgue measure denoted by dx. For example if $V = \mathbb{R}^n$ then there is a fixed symmetric $n\times n$-matrix T with $<u,v>_\gamma = <Tu,v>$ for u,v in $\mathbb{R}^n$ and then

$$\gamma(A) = \int_A (\det T)^{\frac{1}{2}}(2\pi)^{-n/2}\exp(-\tfrac{1}{2}<Tx,x>)dx$$

for the Lebesgue measure dx of $\mathbb{R}^n$. A possibly degenerate Gaussian measure on V is the image of a non-degenerate one on a subspace V_0 of V by the inclusion $\theta:V_0 \to V$. Thus there is a correspondence between Gaussian measures on V and linear subspaces V_0 of V with a given inner product. Strictly speaking translates of these measures are usually also considered to be Gaussian measures.

For a real Banach space E (or a real locally convex topological vector space, ℓ.c.t.v.s) a measure γ on E is said to be Gaussian if for all continuous linear maps $S:E \to V_S$ into some finite dimensional vector space V_S the induced measure

$$\gamma_S \equiv S(\gamma)$$

is Gaussian in the above sense.

Note that by (1) if $f:E \to \mathbb{C}$ factorizes through such an S, so that $f = F \circ S$

$$E \xrightarrow{S} V_S \xrightarrow{F} \mathbb{C}$$

for some measurable F, then

$$\int_E f \, d\gamma = \int_{V_S} F \, d\gamma_S . \tag{2}$$

Such f are called *tame functions*.

The basic structure theorem for Gaussian measures γ is that given one on a separable Banach space there exists a continuous linear injective map

$$i:H \to E$$

of a Hilbert space H, $\langle \, , \, \rangle_H$ into E such that for $S:E \to V_S$ as above the measure γ_S on V_S is associated to the inner product induced on V_S by

$$S \circ i : H \to V_S$$

(with the obvious modification if γ_S is degenerate). For a proof (valid in greater generality) see (12). The triple (i,H,E), or map $i:H \to E$, is called an abstract Wiener space following L. Gross. It completely determines γ, ($\langle\ ,\ \rangle_H$ being fixed). See (26) for detailed discussions and references.

Note that the integral of a cylinder set function $f = F \circ S$ is entirely determined by the restriction $f|H$ of f to the image of H in E (usually identified with H, but, if $\dim E = \infty$, its topology as a subset of E is *not* the same as its Hilbert space topology). In physicist's notation

$$\int_E f d\gamma = \text{"}\int \mathcal{D}[x]\, e^{-\frac{1}{2}|x|_H^2} f[x]\text{"} .$$

Also if $f:E \to \mathbb{C}$ is continuous and bounded and $P_n:E \to P_nH$, $n = 1,2,\ldots,$ is a sequence of projections onto finite dimensional subspaces P_nH of H with $iP_nx \to x$ as $n \to \infty$ for all $x \in E$ then

$$\int_E f\, d\gamma = \lim_{n\to\infty} \int_{P_nH} f|P_nH\, d\gamma_{P_n} \tag{3}$$

by the dominated convergence theorem.

Example: Wiener Measure. For $\tau > 0$ set

$$H = L_0^{2,1} \equiv L_0^{2,1}([0,\tau];\mathbb{R}^n) ,$$

the Hilbert space of paths $\sigma:[0,\tau] \to \mathbb{R}^n$ with

$$\sigma(t) = \int_0^t \rho(s)ds \qquad 0 \le t \le \tau$$

for some ρ in L^2, (so that $\rho = \dot{\sigma}$), where

$$\langle\sigma_1,\sigma_2\rangle_H = \int_0^\tau \langle\dot{\sigma}_1(t),\dot{\sigma}_2(t)\rangle dt \quad .$$

Let $E = C_0 \equiv C_0([0,\tau];\mathbb{R}^n)$ be the space of continuous $\sigma:[0,\tau] \to \mathbb{R}^n$ with $\sigma(0) = 0$, given the norm

$$||\sigma|| = \sup\{|\sigma(t)|:0 \le t \le \tau\} \quad .$$

Then $i:H \to E$, the natural inclusion of H in E, is an abstract Wiener space with corresponding Gaussian measure on E the unique measure w such that for all $0 = t_0<t_1<\ldots<t_m \le \tau$ and Borel sets $A_1,\ldots,A_m$ of $\mathbb{R}^n$

$$w(\{\sigma \in C_0:\sigma(t_j) \in A_j \text{ for } j = 1 \text{ to } m\})$$

$$= \int_{A_1\times\ldots\times A_m} p(t_0,x_0;t_1,x_1)\ldots p(t_{m-1},x_{m-1};t_m,x_m)dx_1\ldots dx_m \tag{4}$$

where $$p(s,x;t,y) = (2\pi(t-s))^{-n/2}\exp(-\frac{|x-y|^2}{2(t-s)}) \quad 0\le s<t \tag{5}$$

and $x_0 = 0 \in \mathbb{R}^n$. The existence of this measure on C_0 is non-trivial, though now standard. In a description which I heard from L. Streit the left hand side of (4) represents the probability of a successful slalom, or the measure of the set of paths which complete a successful slalom. A very successful slalom in fact, since the timing also has to be perfect. The right hand side of (4) happens to be the probability that a

Brownian particle starting from 0 in $\mathbb{R}^n$ executes such a successful slalom according to the Einstein-Smulochowski model of Brownian motion.

Note that in physicist's notation

$$\int_E f\, dw = \int \mathcal{D}[\sigma]\, e^{-\frac{1}{2}\int_0^\tau |\dot{\sigma}(s)|^2 ds} f[\sigma]\,. \tag{6}$$

If we define $S \equiv \rho_{(t_1,\ldots,t_m)} : C_0 \to \mathbb{R}^n \times \ldots \times \mathbb{R}^n = V_S$ by

$$S(\sigma) = (\sigma(t_1),\ldots,\sigma(t_m))$$

then (4) can be written as

$$S(w)(A_1 \ldots A_m) = \int_{A_1\times\ldots\times A_m} p(t_0,x_0;t_1,x_1)\ldots$$

$$p(t_{m-1},x_{m-1};t_m,x_m)dx_1\ldots dx_m$$

or equivalently

$$dw_S = p(t_0,x_0;t_1,x_1)\ldots p(t_{m-1},x_{m-1};t_m,x_m)dx_1\ldots dx_m\quad . \tag{7}$$

For example taking $m = 2$

$$\int_{C_0} |\sigma(t_2)-\sigma(t_1)|^2 dw(\sigma) = \int_{\mathbb{R}^n\times\mathbb{R}^n} |x_2-x_1|^2 dw_S(x_1,x_2) = n|t_2-t_1|. \tag{8}$$

2.3 Cylinder set measures

A *cylinder set measure,* c.s.m., $\{\mu_S\}_S$ on a Banach space, or ℓ.c.t.v.s., E is an assignment of a measure μ_S on V_S for each continuous linear $S:E \to V_S$ into a finite dimensional space V_S , subject to the compatibility condition that if $T:E \to V_T$ and $\rho_{TS}:V_S \to V_T$ are linear with $\rho_{TS} \circ S = T$

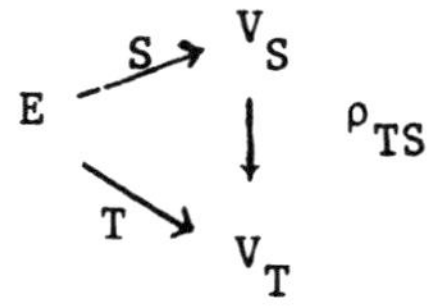

then $\rho_{TS}(\mu_S) = \mu_T$ if $\dim V_T < \infty$.

Examples: (a) a measure μ on E determines a c.s.m. $\{\mu_S\}_S$ on E by $\mu_S = S(\mu)$. Thus some cylinder set measures 'are measures'.

(b) The *canonical Gaussian c.s.m.* $\{\gamma_S\}_S$ on a Hilbert space H , $< \ >_H$ is defined by insisting that each γ_S on V_S , for S surjective, is that Gaussian measure obtained from the inner product induced on V_S by $S:H \to V_S$. When $\dim H = \infty$ this c.s.m. is not a measure in the sense of (a), as the example below shows.

Note that any tame function $f = F \circ S$ on E can be 'integrated' with respect to a c.s.m. $\{\mu_T\}_T$ on E : we can define the integral by

$$\int_E f \, d\{\mu_T\}_T = \int_{V_S} F \, d\mu_S$$

(assuming F is integrable with respect to μ_S of course). The compatibility condition ensures that the result is independent of the representation $f = F \circ S$ of f . One way to proceed further with an integration theory is to try to approximate functions by tame functions: take a sequence of projections $\{P_n : n = 1 \text{ to } \infty\}$ in E with finite dimensional ranges and with $P_n x \to x$ for all x in E, (or take the net of all finite rank projections, or all orthogonal finite rank projections if E is a Hilbert space) and define, for $f: E \to \mathbb{C}$,

$$\begin{aligned}\int_E f \, d\{\mu_T\}_T &= \lim_{n\to\infty} \int_E f\circ P_n \, d\{\mu_T\}_T \\ &= \lim_{n\to\infty} \int_{P_n E} f|P_n E \, d\mu_{P_n}\end{aligned}$$

whenever the integrals and limit exist. See (26) and the references to Gross there. This was one of the first approaches to 'integration' over infinite dimensional Hilbert spaces, see (29), and extends the 'time slicing' approach to Feynmann path integration as we shall see in §2.6. The main problem is that there seems no general way of deciding for which functions the limit exists; another problem is raised by the following example taken from Gross:

Example. Take the canonical Gaussian c.s.m. $\{\gamma_S\}_S$ on a separable Hilbert space H, $\dim H = \infty$. Take an orthonormal base for H and let P_n be the orthogonal projection of H onto the subspace H_n spanned by the first n basis elements. For given $\varepsilon > 0$ let $f: H \to \mathbb{R}$ be given by

$$f(x) = \exp(-\tfrac{1}{2}\varepsilon|x|_H^2) \quad .$$

Then $$\int_H f \circ P_n \, d\{\gamma_S\}_S = \int_{H_n} \exp(-\tfrac{1}{2}\varepsilon|x|^2_{H_n}) d\gamma_{P_n}(x)$$

$$= (2\pi)^{-n/2}\int_{\mathbb{R}^n} \exp(-\tfrac{1}{2}\varepsilon|x|^2)\exp(-\tfrac{1}{2}|x|^2)dx$$

$$= (1+\varepsilon)^{-\frac{1}{2}n}$$

$$\to 0 \quad \text{as} \quad n \to \infty \ .$$

Thus, although f is strictly positive on H and is bounded and continuous its 'integral' over the cylinder set measure would be 0 if we used the approximation procedure as definition. By the dominated convergence theorem this could not happen if the cylinder set measure were a measure. Note also that the physicist's notation

$$\int \mathcal{D}[x] \, e^{-\frac{1}{2}|x|^2_H} e^{-\frac{1}{2}\varepsilon|x|^2_H}$$

for this 'integral' is ambiguous: it could be 1 or 0, 'depending on the normalization' I suppose (but this 'normalization' has to be applied to the approximations in order to turn 0 into 1).

2.4 Radonification

Given a c.s.m. $\{\mu_S\}_S$ on E a continuous linear $T:E \to G$ induces a c.s.m. $\{T(\mu)_S\}_S$ on the Banach space (or ℓ.c.t.v.s) G by

$$T(\mu)_S = \mu_{S \circ T}$$

$$E \xrightarrow{T} G \xrightarrow{S} V_S \ .$$

If this is a measure on G then T is said to *radonify* $\{\mu_S\}_S$ (at least if G is a separable Banach space). This clearly happens if $\dim T(E) < \infty$ but can happen more generally:

Examples: (i) For an abstract Wiener space $i:H \to E$ the map $T = i$ radonifies the canonical Gaussian c.s.m. of H into the Gaussian measure of the abstract Wiener space on E . This follows directly from the definitions.

(ii) If E is a Hilbert space then $T:H \to E$ radonifies the canonical Gaussian c.s.m. of H , and so determines an abstract Wiener space, if and only if T is Hilbert-Schmidt (i.e. trace $T^*T < \infty$). For a proof see (26).

Once a c.s.m. has been radonified by $T:E \to G$ say, there is no problem about integrating functions on E of the form $f = F\circ T$ (i.e. those which extend over G if T is injective and we consider it as an inclusion) where $F:G \to \mathbb{C}$ is integrable with respect to the measure on G , e.g. if F is bounded and continuous when the measure is finite. Thus with an abstract Wiener space $i:H \to E$ one can use the Gaussian measure on E rather than the c.s.m. on H and then apply all the scholarship of measure theory, built up over several decades, with complete confidence. This is at the cost of integrating only functions which have been defined on E . On the other hand Gross has shown that if $i:H \to E$ is an abstract Wiener space then i factorizes

$$\begin{array}{ccc} H & \xrightarrow{i} & E \\ & \searrow_{i_1} & \uparrow k \\ & & E_1 \end{array} \qquad i = k\, i_1$$

where i_1 also radonifies and $k: E_1 \to E$ is a compact linear inclusion. This means that there is a choice of radonifications, and no preferred one, unless external circumstances suggest one. For example the radonification usually chosen for the canonical Gaussian c.s.m. of $H = L_0^{2,1}([0,\tau];\mathbb{R}^n)$ is the inclusion into the space of continuous paths as in §2.2, but the space of Hölder continuous paths exponent $\alpha < \frac{1}{2}$ would be equally good. For the canonical Gaussian c.s.m. on $H^0 = L^2(\mathbb{R}^m)$ it is common to radonify by the inclusion $i: H^0 \to \mathcal{S}(\mathbb{R}^m)'$ of H^0 into the ℓ.c.t.v.s. of tempered distributions: so that the Gaussian measure is considered as lying on $\mathcal{S}(\mathbb{R}^m)'$.

The basic point is that for an abstract Wiener Space $i: H \to E$ the canonical c.s.m. of H controls everything and in some sense E just puts flesh on it. On the other hand $i(H)$ has measure zero as can be seen by the last example in §2.3. The measure on E is quasi-invariant under translation by x if and only if $x \in i(H)$, see (26) or (29) for this generalized 'Cameron-Martin formula'.

2.5 Feynman-Kac Formula

Assume $V: \mathbb{R}^n \to \mathbb{R}$ is continuous and bounded above and $g: \mathbb{R}^n \to \mathbb{C}$ is bounded and measurable. Consider the diffusion equation for $f_t: \mathbb{R}^n \to \mathbb{C}$, $t \geq 0$:

$$\begin{aligned} \frac{\partial f_t}{\partial t} &= \tfrac{1}{2}\Delta f_t + V f_t \qquad t > 0 \\ f_0 &= g \ . \end{aligned} \tag{D.E.}$$

For $t \geq 0$ define $P_t g: \mathbb{R}^n \to \mathbb{C}$ by

$$P_t g(x_0) = \int_{C_0} e^{\int_0^t V(x_0+\sigma(s))ds} g(x_0+\sigma(t))dw(\sigma) \tag{9}$$

for Wiener measure w on $C_0 \equiv C_0([0,t]; \mathbb{R}^n)$. Then

(i) $P_t(P_s g) = P_{t+s} g$

(ii) $\lim_{t \downarrow 0} \frac{1}{t} (P_t g(x_0) - g(x_0)) = \tfrac{1}{2}\Delta g(x_0) + V(x_0) g(x_0)$

for all $x_0 \in \mathbb{R}^n$ whenever g is C^2 with bounded first and second derivatives.

(iii) $P_t g$ is a classical solution to (D.E.) provided V is sufficiently smooth.

Proof We will sketch the proofs of (i) and (ii) only: for more, and results about more general V see (30).

(i) Observe that as a Hilbert space $L_0^{2,1}([0,t+s];\mathbb{R}^n)$ is naturally isometrically isomorphic to the direct sum $L_0^{2,1}([0,s]; \mathbb{R}^n) \oplus L_0^{2,1}([s,t+s]; \mathbb{R}^n)$ the second factor of which is in turn naturally isometrically isomorphic to $L^{2,1}([0,t];\mathbb{R}^n)$. This splitting is reflected in a corresponding decomposition of the space of continuous paths and of the Wiener measure into a product of Wiener measures. Symbolically:

$$\begin{array}{ccccc} C_0([0,t+s];\mathbb{R}^n) & & C_0([0,s];\mathbb{R}^n) & & C_0([0,t];\mathbb{R}^n) \\ \uparrow & \approx & \uparrow & \oplus & \uparrow \\ L_0^{2,1}([0,t+s];\mathbb{R}^n) & & L_0^{2,1}([0,s];\mathbb{R}^n) & & L_0^{2,1}([0,t];\mathbb{R}^n) \end{array}$$

The semigroup property (i) is then an immediate consequence of Fubini's theorem.

(ii) *Case* (a): $V \equiv 0$.

$$P_t g(x_0) = \int_{C_0} g(x_0+\sigma(t))dw(\sigma)$$

$$= \int_{\mathbb{R}^n} (2\pi t)^{-n/2} \exp(-\frac{|x|^2}{2t}) g(x_0+x)dx$$

(we are integrating a tame function)

$$= \int_{\mathbb{R}^n} p(0,x_0;t,y)g(y)dy$$

and p can be recognized to be the fundamental solution of the heat equation $\frac{\partial f_t}{\partial t} = \frac{1}{2}\Delta f_t$, i.e. the 'Euclidean propagator'.

Case (b) V continuous.

$$\lim_{t\downarrow 0} \frac{1}{t}(P_t g(x_0)-g(x_0)) = \lim_{t\downarrow 0}\int_{C_0} \frac{1}{t}\{e^{\int_0^t V(x_0+\sigma(s))ds} -1\}g(x_0+\sigma(t))dw(\sigma)$$

$$+ \lim_{t\downarrow 0} \frac{1}{t}\{\int_{C_0} g(x_0+\sigma(t))dw(\sigma)-g(x_0)\}$$

$$= V(x_0)g(x_0) + \frac{1}{2}\Delta g(x_0)$$

by the dominated convergence theorem and case (a).

2.6 Time Slicing

Let $\Pi = \{t_0, \ldots, t_m\}$ be a partition of $[0,t]$ with $0 = t_0 < t_1 < \ldots < t_m = t$. Define the corresponding piecewise linear approximation

$$P_\pi \sigma : [0,t] \to \mathbb{R}^n$$

to a path $\sigma : [0,t] \to \mathbb{R}^n$

by
$$P_\pi(\sigma)(s) = \sigma(t_{j-1}) + (s - t_{j-1})(\Delta_j t)^{-1} \Delta_j \sigma$$
$$t_{j-1} \le s \le t_j$$

where
$$\Delta_j \sigma = \sigma(t_j) - \sigma(t_{j-1}) \ , \ \Delta_j t = t_j - t_{j-1} \ .$$

Then $P_\pi \sigma \to \sigma$ uniformly on $[0,t]$ for each continuous σ as mesh $\Pi \to 0$. Also each P_π can be considered as an orthogonal projection on $L_0^{2,1}([0,t];\mathbb{R}^n)$ with finite rank and which extends to a continuous linear map still written P_π, from C_0 to $L_0^{2,1}$. In particular when g and V are continuous with g bounded and V bounded above we can apply equations (3) and (7) to the definition of $P_t g(x_0)$ above to obtain, for $z = -i$,

$$P_t g(x_0) = \lim I_\pi^z \quad \text{as mesh } \Pi \to 0$$

where
$$I_\pi^z = \int_{\mathbb{R}^n \times \ldots \times \mathbb{R}^n} \left\{ \prod_{j=1}^{m} (2\pi i z \Delta_j t)^{-\frac{1}{2}n} \, e^{\frac{i}{z} A(t_{j-1}, \rho_{j-1}; t_j, \rho_j)} \right. \\ g(\rho_m) d\rho_1 \ldots d\rho_m \qquad (10)$$

for $A(t_{j-1},x;t_j,y) = \frac{1}{2}\frac{|x-y|^2}{t_j} - \int_{t_{j-1}}^{t_j} V(x+(s-t_{j-1})(\Delta_j t)^{-1}(y-x))ds$

$x,y \in \mathbb{R}^n$.

3. FEYNMAN PATH INTEGRALS

3.1 Oscillatory Integrals and Fresnal Integrals

The approach adopted here is taken from (16), (17), (18). The object is to make mathematical sense of Feynman's original definition of the path integral for a non-relativistic quantum mechanical particle (essentially $\lim I_\pi^z$ as in §2.6 but with $z = 1$). The first problem is to give meaning to

$$(2\pi i)^{-m/2}\int_{\mathbb{R}^m} f(x)\, e^{\frac{1}{2}i|x|^2}\, dx .$$

When f is Lebesgue integrable there is no problem, but if for example $f(x) = x^2$ some regularization procedure must be laid down. For this it is convenient to follow Hörmander and use *oscillatory integrals*. It will also be useful to fix a symmetric linear map $B:\mathbb{R}^m \to \mathbb{R}^m$ and set $T = 1+B$. Suppose $f:\mathbb{R}^m \to \mathbb{C}$ is measurable. Then for $\varepsilon > 0$ and an element $\phi \in S(\mathbb{R}^m)$ with $\phi(0) = 1$ of the Schwartz space of rapidly decreasing functions $\phi:\mathbb{R}^m \to \mathbb{C}$, (27), set

$$\ell_\phi^\varepsilon(f) = \int_{\mathbb{R}^m} (2\pi i)^{-m/2}\phi(\varepsilon x)e^{i/2\langle Tx,x\rangle}\, dx$$

and $\ell_\phi(f) = \lim_{\varepsilon \downarrow 0} \ell_\phi^\varepsilon(f)$ whenever the integrals and limit exist.

When $\ell_\phi(f)$ exists and is independent of ϕ we will write

$f \in I_B(\mathbb{R}^m;\mathbb{C})$ and define the *oscillatory integral* by

$$\int^{0} (2\pi i)^{-m/2} f(x)e^{i/2\langle Tx,x\rangle} dx = \ell_\phi(f) .$$

Exercise: Compute $\int^{0}_{\mathbb{R}} x^p (2\pi i)^{-\frac{1}{2}} e^{i/2|x|^2} dx \quad p = 1,2..$

(Hint: integrate the expression for $\ell^\varepsilon_\phi(f)$ by parts.)

These oscillatory integrals are related to Fresnel integrals: for a Hilbert space H , e.g. $H = \mathbb{R}^m$, define $F(H)$ to be the space of those $f:H \to \mathbb{C}$ which are the Fourier transform of some complex measure μ_f

$$f(\alpha) = \int_H e^{i\langle\alpha,x\rangle_H} d\mu_f(x) \qquad \alpha \in H .$$

Then the basic lemma of (2) when $B = 0$, and (17) (18) in general, is that if $T = 1+B$ is as above and is an isomorphism then

$$f \in F(\mathbb{R}^m) \Rightarrow f \in I_B(\mathbb{R}^m;\mathbb{C})$$

and

$$(2\pi i)^{-m/2} \int^{0}_{\mathbb{R}^m} f(x)e^{i/2\langle Tx,x\rangle} dx$$

$$= \exp\{-\tfrac{\pi i}{2}\text{Ind}T\}|\det T|^{-\frac{1}{2}} \int_{\mathbb{R}^m} e^{-\frac{1}{2}i\langle T^{-1}x,x\rangle} d\mu_f(x) \quad (11)$$

where the (Maslov) index, Ind T, is the total dimension of the negative eigenspaces of T .

3.2 Feynman Maps

In Feynman path integration it is customary and convenient

to consider paths ending at a given point, rather than starting there, as is convenient for Wiener integration. For $\sigma:[0,t] \to \mathbb{R}^n$ define

$$J(\sigma):[0,t] \to \mathbb{R}^n$$

by $$J(\sigma)(s) = \sigma(t-s)$$

and now take H to be the Hilbert space

$$H \equiv {}_0L^{2,1}([0,t];\mathbb{R}^n) = \{J(\sigma):\sigma \in L_0^{2,1}([0,t];\mathbb{R}^n)\}$$

with inner product so that $J : {}_0L^{2,1} \to L_0^{2,1}$ is isometric. Take Π and P_π as in §2.6. For non-zero $z \in \mathbb{C}$ with positive imaginary part, and for suitable $f:H \to \mathbb{C}$ define

$$F_\pi^z(f) \in \mathbb{C}$$

by $$F_\pi^z(f) = (2\pi iz)^{-\frac{1}{2}nm}(\Delta_1 t)^{-\frac{1}{2}n}\dots(\Delta_m t)^{-\frac{1}{2}n}\int_{P_\pi H}^{0} f(x)e^{\frac{i}{2z}|x|^2}dx .$$

Define $$F^z(f) = \lim_{\text{mesh}\Pi\to 0} F_\pi^z(f)$$

when it exists. For $z = 1$ this is the *Feynman integral* of f, to be written as $F(f)$. The basic results are as follows (17), (18) :

Case (i), $z = -i$: If $f\circ J$ has a continuous bounded extension

$$f\tilde{\circ}J : C_0([0,t]; \mathbb{R}^n) \to \mathbb{C}$$

then $$F^{-i}(f) = \int_{C_0} f\tilde{\circ}J(\sigma)dw(\sigma) .$$

Case (ii), $z = 1$: Let $f \in \mathcal{F}(H)$ and let $B: H \to H$ be continuous linear, of trace class (= nuclear), with $B^* = B$ and $T = 1+B : H \to H$ an isomorphism. Set

$$g(\sigma) = \exp(\tfrac{1}{2}i\langle B\sigma,\sigma\rangle_H) f(\sigma) \qquad \sigma \in H .$$

Then $\mathcal{F}(g)$ exists and

$$\mathcal{F}(g) = \exp(-\tfrac{1}{2}\pi i \operatorname{Ind}T)\,|\det T|^{-\frac{1}{2}} \int_H e^{-\frac{1}{2}i\langle\sigma,T^{-1}\sigma\rangle_H} d\mu_f(\sigma) .$$

Case (i) is essentially just formula (3) again, while case (ii) follows from the basic lemma in §3.1.

3.3 <u>Feynman path integrals and the Schrödinger equation</u>

Suppose $V_0 \in \mathcal{F}(\mathbb{R}^n)$ and $V(x) = \tfrac{1}{2}x\Omega^2x + V_0(x)$ where $x \mapsto x\Omega^2x$ is a positive definite quadratic form. Let H be the corresponding anharmonic oscillator Hamiltonian

$$H = -\tfrac{1}{2}\Delta + V .$$

Then for $\phi_0 \in \mathcal{F}(\mathbb{R}^n)$ the Feynman-Kac-Itô formula is valid (2), (17), (18) for the solution of the Schrödinger equation

$$\frac{\partial\phi_t}{\partial t} = \tfrac{1}{2}\, i\Delta\phi_t - iV\phi_t ,$$

i.e.
$$\phi_t(x_0) \equiv (e^{-itH}\phi_0)(x_0) = \mathcal{F}\left(e^{-i\int_0^t V(x_0+\sigma(s))ds}\,\phi(x_0+\sigma(0))\right).$$

If $\phi_0 \in L^2(\mathbb{R}^n)$ the formula still holds in the L^2 sense and the convergence of $\mathcal{F}_\pi^1$ as mesh $\Pi \to 0$ is only in L^2 , not for each $x_0 \in \mathbb{R}^n$.

The condition on V_0 is very restrictive. Every $V_0 \in F(\mathbb{R}^n)$ is bounded and uniformly continuous. It must be emphasized that rigorous mathematical results about Feynman path integrals are very limited. Even when it can be proved that the 'integrals' exist and give solutions of the Schrödinger equation it is not at all easy to use them to get information about such solutions in a rigorous manner: for example to obtained semi-classical expansions. However the latter has been done, under slightly more restrictive conditions, for the anharmonic oscillator just mentioned (1). See also (20) where a slightly different 'Feynman integral' is used. Mathematically they seem to be more of a challenge than an effective tool. In quantum physics their use is now standard and occurs in situations of much greater complexity than we have considered. A dip into the proceedings of the Nato summer school on 'Relativity, Groups, Topology' at Les Houches, July 1983, to be published by North-Holland, will convince the mathematician how far ahead physicists have got. For their less esoteric (but still heuristic) use in non-relativistic quantum mechanics see (10).

4. PATH INTEGRATION ON RIEMANNIAN MANIFOLDS

4.1 Wiener measure and rolling without slipping

Let Δ be the Laplace-Beltrami operator, $\Delta = \text{divgrad}$, of an n-dimensional Riemannian manifold M. As for $\mathbb{R}^n$ there is a fundamental solution $p(s,x;t,y)$, $0 \le s < t$, $x,y \in M$, to the heat equation for $f_t : M \to \mathbb{R}$, $t \ge 0$

$$\frac{\partial f_t}{\partial t} = \tfrac{1}{2}\Delta f_t$$

so that $\quad f_t(x) = \int_M p(s,x;t,y) f_s(y) dy$

where the integration is with respect to the volume element of M . For each $x_0 \in M$ it is possible by general theory (e.g. Markov process theory) to define *Wiener measure* w_{x_0} on the space $C_{x_0}([0,t];M)$ of continuous

$$\sigma : [0,t] \to M \quad \text{with} \quad \sigma(0) = x_0$$

by analogy with the $\mathbb{R}^n$ case as the unique measure such that if $A_1, \ldots, A_m$ are Borel sets in M and $0 = t_0 < \ldots < t_m$ then

$$w_{x_0}\{\sigma : \sigma(t_j) \in A_j \quad \text{for} \quad j = 1 \text{ to } m\}$$
$$= \int_{A_1 \times \ldots \times A_m} p(t_0,x_0;t_1,x_1) \ldots p(t_{m-1},x_{m-1};t_m,x_m) \, dx_1 \ldots dx_m \qquad (13)$$

c.f. equation (4). (In some circumstances it is necessary to use the one point compactification of M (14).)

We shall consider a geometrical method of constructing this measure which reveals a general method of going from path integration on flat spaces to path integration on curved spaces. The method is based on the Cartan development, a construction which lies in the roots of differential geometry.

The *Cartan development* can be considered as a map

$$R: L_0^{2,1}([0,t];\mathbb{R}^n) \to L_{x_0}^{2,1}([0,t];M)$$

for M complete. Intuitively, take a copy of $\mathbb{R}^n$ and place M on it with x_0 at the origin (this requires a choice of isometry, or *frame*, $u_0: \mathbb{R}^n \to T_{x_0}M$, where $T_{x_0}M$ denotes the tangent space at x_0 to M). Then given $\sigma:[0,t] \to \mathbb{R}^n$ with $\sigma(0) = 0$, roll M along σ, on the copy of $\mathbb{R}^n$, without slipping. The point of contact of M with $\mathbb{R}^n$ then describes a path $R(\sigma)$ on M which is the Cartan development of σ. It depends on the choice of initial frame u_0 (traditionally σ is a path in $T_{x_0}M$ and u_0 is not required, but it is more convenient for us to do it this way). Note that at each time s in $[0,t]$ we have a frame $u_s: \mathbb{R}^n \to T_{x_s}M$, where $x_s = R(\sigma)(s)$. These frames would be treated most naturally as only affine frames but we take the linear part and let u_s denote the corresponding orthonormal frame (this distinction is ignored in the diagram):

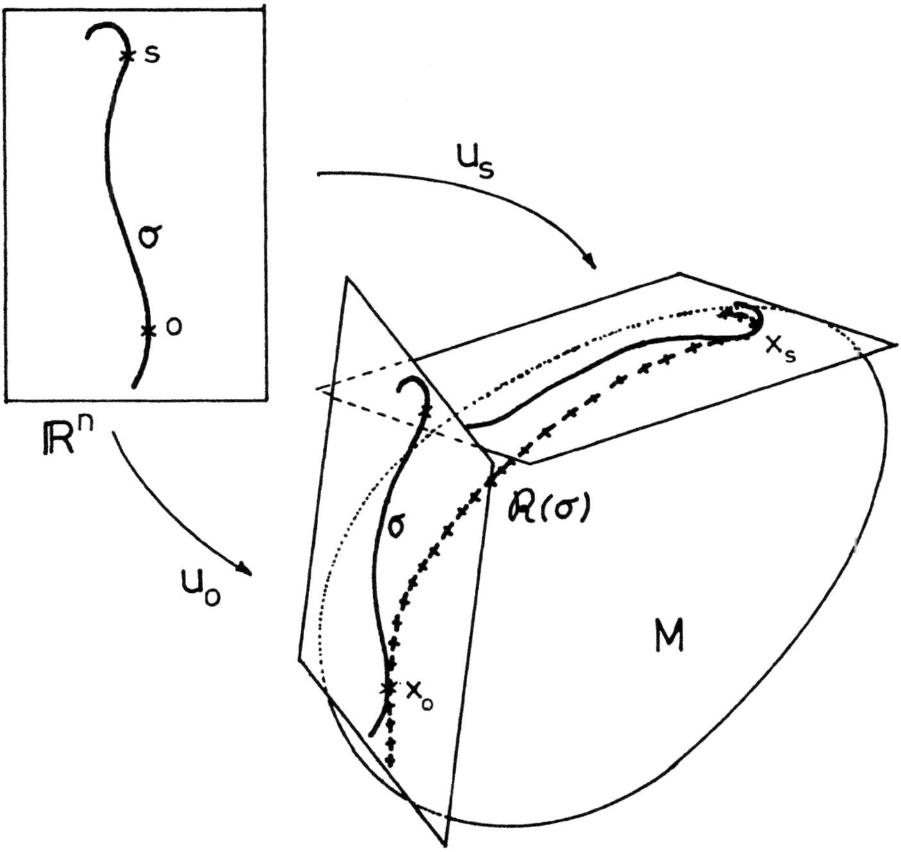

For a mathematical description it is best to work on the bundle $p: O(M) \to M$ of orthonormal frames of M. If $u \in p^{-1}(x)$ we consider u as an isometry $u: \mathbb{R}^n \to T_x M$, (the corresponding 'n-bein' is the set of tangent vectors $u(e_1), \ldots u(e_n)$ for the standard base $e_1, \ldots, e_n$ of $\mathbb{R}^n$). The Levi-Civita connection determines a map

$$h_u : T_x M \to T_u O(M) \qquad u \in O(m) \ , \ x = p(u)$$

which gives the 'horizontal lift' of $v \in T_x M$ over u. In particular

$$T_u p(h_u(v)) = v \ .$$

For each $e \in \mathbb{R}^n$ define a vectorfield $X(\cdot)e$ on $O(M)$ by

$$X(u)e = h_u(u(e)) \in T_u O(M) \ .$$

Then $\sigma:[0,t] \to \mathbb{R}^n$ determines a time dependent vector field

$$(u,s) \to X(u) \frac{d\sigma}{ds}$$

on $O(M)$.

For our given u_0 we can solve

$$\frac{du_s}{ds} = X(u_s) \frac{d\sigma}{ds} \qquad 0 \leq s < t \qquad (14)$$

to obtain $u_s \in O(M)$, $0 \leq s \leq t$, assuming M is complete. Set

$$x_s = p(u_s) \ .$$

Then $x_s = \mathcal{R}(\sigma)(s)$, the development of σ .

Note that we have so far used no special properties of Levi-Civita connections, or Riemannian manifolds. The construction works equally well for any linear connection on a manifold (e.g. the Levi-Civita connection of a Lorentz manifold) and extends to more general bundles than the frame bundle using the notion of *soldering* (25). In the more general situation one can *define* the connection to be complete if solutions to (14) exist for all time (25). In the Riemannian case

$|\dot{\sigma}(s)| = |\mathcal{R}(\dot{\sigma})(s)|$ so that

$$\tfrac{1}{2}\int_0^t |\dot{\sigma}(s)|^2 ds = \tfrac{1}{2}\int_0^t |\dot{x}_s|^2 ds . \tag{15}$$

The idea of the *stochastic development* is to use $\mathcal{R}$ to map the Wiener measure on paths in $\mathbb{R}^n$ to a measure on paths in M. In view of (15) this measure would be written in physicist's notation as

$$\mathcal{D}[x] e^{-\frac{1}{2}\int_0^t |\dot{x}_s|^2 ds}$$

for paths x on M, and should be the required Wiener measure on $C_{x_0}([0,t];M)$. The difficulty is that one can only roll along paths with some differentiability, as can be seen from equation (14), and the set of paths which are even of bounded variation has Wiener measure zero in $C_0([0,t];\mathbb{R}^n)$. However it turns out that one can interpret (14) as a 'stochastic differential equation' the solution of which in some sense gives an extension of $\mathcal{R}$ over C_0. For our purposes it is easiest tc explain this by taking piecewise linear approximations to continuous paths:

Take Π and P_π as in §2.6. For continuous σ and fixed u_0 let $u_s^\pi(\sigma)$, $s \geq 0$, be the solution of

$$\frac{du_s^\pi}{ds}(\sigma) = X(u_s^\pi(\sigma)) \frac{d}{ds}(P_\pi \sigma) \qquad s \geq 0$$

$$u_0^\pi(\sigma) = u_0 .$$

Set $x_s^\pi(\sigma) = p(u_s^\pi(\sigma)) = \mathcal{R}(P_\pi(\sigma))$.

Then, if M is complete with Ricci curvature bounded below, as mesh $\Pi \to 0$ so $u_s^\pi : C_0([0,t]; \mathbb{R}^n) \to O(M)$ converges uniformly in $0 \le s \le t$, in w-measure, to some

$$u_s : C_0([0,t]; \mathbb{R}^n) \to O(M) .$$

Also x_s^π converges similarly to some x_s with $x_s(\sigma) = p(u_s(\sigma))$ and the resulting map

$$\tilde{\mathcal{R}} : C_0([0,t]; \mathbb{R}^n) \to C_{x_0}([0,t];M)$$
$$\tilde{\mathcal{R}}(\sigma)\,(s) = x_s(\sigma)$$

sends Wiener measure w on C_0 to Wiener measure w_{x_0} as defined in §4.1. See (15) for a detailed proof.

The Ricci curvature condition is a condition due to S.T. Yau to ensure that $\tilde{\mathcal{R}}(\sigma)$ does not go out to infinity in M in finite time, (or equivalently that $\int_M p(s,x,t,y)dy = 1$ all $0 \le s < t$ and $x \in M$) , see (14) or (15). This is not implied by completeness alone, but will be an assumption on M from now on.

Note that although $\tilde{\mathcal{R}}$ depends on the choice of the initial frame u_0 the resulting measure $\tilde{\mathcal{R}}(w)$ does not. This is due to the fact that the Wiener measure on $\mathbb{R}^n$ is invariant under orthogonal transformations of $\mathbb{R}^n$ (because the $p(s,x;t,y)$ are invariant). For this it is important that we are using orthonormal frames and have Riemannian metrics. A well defined measure on $C_{u_0}([0,t];O(M))$ is always induced from w but to get one independent of u_0 on $C_{x_0}([0,t];M)$ when the manifold is not Riemannian it would be necessary to replace w by some measure invariant under the relevant structure group of M (but see (11) for the Lorentz case).

There is also the Feynman-Kac formula for the solution $f_t:M \to \mathbb{C}$, $0 \le t < \infty$ to

$$\frac{\partial f_t}{\partial t} = \tfrac{1}{2}\Delta f_t + Vf_t \qquad t > 0$$
$$f_0 = g$$

for suitable $V:M \to \mathbb{C}$ and $g:M \to \mathbb{C}$. As for the case $M = \mathbb{R}^n$ we define

$$P_t g:M \to \mathbb{C}$$

by
$$P_t g(x_0) = \int_{C_{x_0}([0,t];M)} e^{\int_0^t V(x_s)ds} g(x_t)dw_{x_0}(x)$$

so that by (1) $$P_t g(x_0) = \int_{C_0([0,t];\mathbb{R}^n)} e^{\int_0^t V(x_s(\sigma))ds} g(x_t(\sigma))dw(\sigma) .$$

Then for suitable g and V we have $f_t = P_t g$.

4.2 The Pauli-Van-Vleck-DeWitt propagator

For bounded continuous V and g the above formula for $P_t g(x_0)$ can be written as

$$P_t g(x_0) = \lim_{\text{mesh}\Pi\to 0} I_\pi$$

where
$$I_\pi = \int_{C_0([0,t];\mathbb{R}^n)} e^{\int_0^t V(x_s^\pi(\sigma))ds} g(x_t^\pi(\sigma))dw(\sigma)$$

for $x_s^\pi(\sigma) = R(P_\pi(\sigma))(s)$ as before.

Set $k = (2\pi)^{-\frac{1}{2}mn} \prod_{j=1}^{m} (\Delta_j t)^{-\frac{1}{2}n}$

and following (16) make the change of integration variable to $\Delta = x_j - x_{j-1}$. Then

$$I_\pi = k\int_{\mathbb{R}^n\times\ldots\times\mathbb{R}^n} \exp\{-\tfrac{1}{2}\sum_{j=1}^{m} \frac{|\Delta_j|^2}{\Delta_j t} + \int_0^t V(\rho(\Delta_1,\ldots,\Delta_m)(s))ds\}$$
$$g(\rho(\Delta_1,\ldots,\Delta_m)(t))d\Delta_1\ldots d\Delta_m \qquad (15)$$

where $\rho(\Delta_1,\ldots,\Delta_m)$ is the Cartan development of the piecewise linear path σ on $\mathbb{R}^n$ with $\sigma(0) = 0$ and $\Delta_j\sigma = \Delta_j$ for $j = 1$ to m .

If we assume that M has the property that each pair of points x,y in M can be joined by a unique geodesic (e.g. M simply connected, complete, with non-positive curvature) we can simplify I_π further. For this let $A(r,x;s,y)$ be the action along the geodesic γ from x to y parametrized proportional to arc length with $\gamma(r) = x$ and $\gamma(s) = y$:

$$A(r,x;s,y) = \tfrac{1}{2}\int_r^s |\dot{\gamma}(u)|^2 du - \int_r^s V(\gamma(u))du . \tag{16}$$

Let $\qquad U(\Delta_1,\ldots,\Delta_{j-1}):T_{x_0}M \to T_{\rho_{j-1}}M ,$

for $\qquad \rho_{j-1} = \rho(\Delta_1,\ldots,\Delta_{j-1})(t_{j-1}) ,$

be the isometry obtained by parallel translation along $\rho(\Delta_1,\ldots,\Delta_{j-1})$. Then using the exponential map of M and setting $\tilde{\Delta}_j = u_0\Delta_j$:

$$\rho(\Delta_1,\ldots,\Delta_m)(s) = \exp_{x_0} s\tilde{\Delta}_1 \qquad\qquad 0 \le s \le t_1$$

and more generally, for $t_{j-1} \le s \le t_j$

$$\rho(\Delta_1,\ldots,\Delta_m)(s) = \exp_{\rho_{j-1}}\left(\frac{s-t_{j-1}}{\Delta_j t}\, U(\Delta_1,\ldots,\Delta_{j-1})\tilde{\Delta}_j\right) .$$

Our assumption implies that $\exp_{x_0}:T_{x_0}M \to M$ is a diffeomorphism so we can change integration variables again, (see (16)), to get $I_\pi = I_\pi^z$ for $z = -i$ and

$$I_\pi^z = (2\pi i z)^{-\frac{1}{2}mn}\int_{M\times\ldots\times M} \prod_{j=1}^m e^{\frac{i}{z}A(t_{j-1},\rho_{j-1};t_j\rho_j)} \cdot\cdot$$

$$(\Delta_j t)^{n/2}\mathcal{D}(t_{j-1},\rho_{j-1};t_j,\rho_j)\ g(\rho_m)d\rho_1\ldots d\rho_m \tag{17}$$

where $\mathcal{D}(t_{j-1},a;t_j,b) = (\Delta_j t)^{-n}|\det D(\exp_a^{-1})(b)|$ $\quad a,b \in M$ is the 'Van-Vleck determinant', $(\Delta_j t)^{-n}$ times the Jacobian determinant of the inverse of the exponential map.

Thus under our hypotheses when $z = -i$ the limit of I_{π}^{z} as mesh $\Pi \to 0$ gives the solution of our diffusion. For $z = 1$ our expression is very close to the 'Pauli-Van-Vleck-DeWitt propagator' which was proposed as the time slicing approximation to the Feynman path integral of a non-relativistic quantum particle with a Riemannian manifold as configuration space. The difference is that we have $(\Delta_j t)^{n/2}\mathcal{D}$ where they would have $\mathcal{D}^{\frac{1}{2}}$. However, as DeWitt observed, theirs would be expected to lead to solutions of the Schrödinger equation

$$\frac{\partial\phi_t}{\partial t} = \tfrac{1}{2}\, i\Delta\phi_t - iV\phi_t - \frac{1}{12}\, i\rho$$

where ρ is the scalar curvature of M. See (9) for bibliography and discussion. In any case I know of no rigorous results about the convergence of the Feynman sums, i.e. of I_{π}^{z} for $z = 1$, except on flat manifolds.

5. GAUGE INVARIANT EQUATIONS; DIFFUSION AND DIFFERENTIAL FORMS

5.1 Quantum particle in a classical magnetic field

Let the particle have configuration space our Riemannian manifold M and suppose that the electromagnetic potential on M comes from a connection on a principal bundle $P:G(B) \to BM$ with structure group $G = U(1)$ or $\mathbb{R}$ and that the wave function of our particle is a section ψ of an associated complex line bundle $\tilde{P}:B \to M$.

Any smooth, or even $L^{2,1}$, path x on M has a unique horizontal lift to a path $\tilde{x}$ on $G(B)$ with given initial point $\tilde{x}_0 \in P^{-1}(x_0)$

$$\begin{array}{ccc} & & G(B) \\ & \overset{\tilde{x}}{\nearrow} & \downarrow P \\ [0,t] & \xrightarrow{x} & M \end{array} .$$

If $G(B) = M\times G$ then $\tilde{x}_t = (x_t, g_t)$ where

$$\frac{dg_t}{dt} = -i\ \Gamma(x_t)\left(\frac{dx_t}{dt}\right)g_t \tag{18}$$

for Γ the Christoffel symbol of the connection of $G(B)$, so that

$$\Gamma = \frac{eA}{\hbar c} : TM \to \mathbb{R}$$

considered as one-forms on M for A the vector potential in our trivialization. Parallel translation b_t along x of an element $b_0 \in B$ over x_0 is then given by $b_t = g_t b_0$ when $g_0 = 1$, or by

$$b_t = \tilde{x}_t \tilde{x}_0^{-1} b_0$$

when we consider the elements $\tilde{x}_t$ of $G(B)$ as isomorphisms

$$\tilde{x}_t : \mathbb{C} \to \tilde{P}^{-1}(x_t) .$$

To write down a path integral solution for the 'Euclidean' or 'imaginary time' Schrödinger equation for our wave function we will need to parallel translate along the paths $\mathcal{R}(\sigma)$ on M. Since these are not differentiable, equation (18) can not be applied directly. However for $x = \mathcal{R}(\sigma)$ it can be rewritten as

$$\frac{dg_s(\sigma)}{ds} = -i\ \Gamma(x_s(\sigma))(u_s(\sigma)\frac{d\sigma(s)}{ds})g_s(\sigma) \tag{19}$$

for $\tilde{x}_s(\sigma) = (x_s(\sigma), g_s(\sigma))$ the lift of $x_s(\sigma)$, $0 \le s \le t$, and then interpreted as a (Stratonovich) stochastic differential equation with solution

$$g_s(\sigma) = \exp\{-i\int_0^s \Gamma(x_r(\sigma))(u_r(\sigma)\circ d\sigma_r)\}g_0 \tag{20}$$

in terms of a Stratonovich integral. In fact this solution is just the limit in w-measure as mesh $\Pi \to 0$ of the solutions $g_s(P_\pi\sigma)$ for the piecewise linear approximations $P_\pi\sigma$.

Let $w^{\hbar}$ be the modification to Wiener measure on $C_0([0,t];\mathbb{R}^n)$ obtained by using the inner product $\frac{1}{2}\int_0^t \hbar^{-1}|\dot{\sigma}(s)|^2 ds$ on $L_0^{2,1}$. This corresponds to replacing our basic heat equation on $\mathbb{R}^n$ by $\frac{\partial f_t}{\partial t} = \frac{1}{2}\hbar\Delta f_t$. Let Δ_P denote the covariant Laplacian on sections ψ of B, corresponding to the covariant differentiation

$$\psi \to d\psi + \frac{ie}{\hbar c}A .$$

Then under standard conditions the solution $\psi_t : M \to B$, $t \ge 0$, to

$$\frac{\partial\psi_t}{\partial t} = \tfrac{1}{2}\hbar\,\Delta_P\psi_t + \frac{V}{\hbar}\psi_t \tag{21}$$

is $$\psi_t(x_0) = \int_{C_0([0,t];\mathbb{R}^1)} \tilde{x}_0\tilde{x}_t(\sigma)^{-1} e^{\frac{1}{\hbar}\int_0^t V(x_s(\sigma))ds}\,\psi_0(x_t(\sigma))\,dw^{\hbar}(\sigma) \tag{22}$$

so that given a trivialization of $G(B)$

$$\psi_t(x_0) = \int_{C_0([0,t];\mathbb{R}^n)} e^{\frac{ie}{\hbar c}\int_0^t A(x_s(\sigma))\circ dx_s(\sigma) + \frac{1}{\hbar}\int_0^t V(x_s(\sigma))ds}\,\psi_0(x_t(\sigma))\,dw^{\hbar}(\sigma) \tag{23}$$

See (9), (15) and the references therein, and (30) for the case $M = \mathbb{R}^n$.

The analogous equation for a particle in a non-abelian gauge field can be handled similarly with solution given by the direct analogue of equation (22). However in this case there is no simple formula like (20) for the solution to the corresponding equation (19) since this is now a matrix equation (but see (5) for a probabilistic formula in a similar situation, and also for particles with spin).

5.2 Heat equation for differential forms

There is an analogous formula to (23) for the solution to the heat equation

$$\frac{\partial\phi_t}{\partial t} = \tfrac{1}{2}\Delta\phi_t \tag{24}$$

for differential forms on M , with ϕ_0 given. In this case the principal bundle is just the frame bundle O(M) and we already have the 'horizontal lift' of $x_s(\sigma)$ to O(M) : it is just $u_s(\sigma)$. However the Wietzenböck formula, for a 1-form ϕ ,

$$\Delta\phi = \text{trace } \nabla^2\phi - \text{Ric}(-,\phi^{\#})$$

where $\phi^{\#}$ is the vector field corresponding to ϕ (i.e. obtained by raising the indices), shows that this Laplacian is not just the 'covariant Laplacian' trace ∇^2 . To take the Ricci curvature term into account define the vector fields $v_s(\sigma)$ along $x_s(\sigma)$, $s \geq 0$, i.e. the lifts of $x_s(\sigma)$ to TM , by

$$\frac{D}{\partial s} v_s(\sigma) = -\tfrac{1}{2} \text{Ric}(v_s(\sigma),-)^{\#} \qquad (25)$$

by which we mean $v_s(\sigma) = u_s(\sigma)\hat{v}_s(\sigma)$ where

$$\frac{d}{ds}(\hat{v}_s(\sigma)) = -\tfrac{1}{2}u_s(\sigma)^{-1} \text{Ric}(u_s(\sigma)\hat{v}_s(\sigma),-)^{\#} \qquad (26)$$

for given $v_0 \in T_{x_0}M$. The solution ϕ_t to (24) is then given by

$$\phi_t(v_0) = \int_{C_0([0,t];\mathbb{R}^n)} \phi_0(v_s(\sigma))dw(\sigma) \quad . \qquad (27)$$

See (15) for a detailed discussion. The formula for p-forms is similar although the curvature term which comes in via the Weitzenbock formula is more complicated if $p > 1$, (22).

ACKNOWLEDGEMENTS

The material on Wiener integration etc., was mostly learnt or worked out in collaboration with J. Eells, and that on Feynman integration with A. Truman. The latter being helped by an S.E.R.C. grant (GR/C/13644). Participation in the Les Houches summer school / NATO A.S.I. on "Relativity, Groups, and Topology" July 1983 provided stimulation and material for the introduction: in particular B. DeWitt allowed me a preview of his monograph, R. Warner led me to several of the references, and B. Zumino sketched the supersymmetric path integral approach to the index theorem for the Dirac operator.

The diagram was prepared at the Hochschule der Bundeswehr, Hamburg, and the typing is by Peta McAllister, Mathematics department, University of Warwick.

REFERENCES

The bibliographies in references (1), (2), (8), (10), (11), (15), (26), (30) below may be found particularly useful for additional information.

1. Albeverio, S., Blanchard, Ph. & Høegh-Krohn, R. (1982). *Feynman path integrals and the trace formula for the Schrödinger operators.* Comm. Math. Phys., 83, no.1, pp. 49-76.

2. Albeverio, S.A. & Høegh-Krohn, R.J. (1976). *Mathematical Theory of Feynman Path Integrals.* Lecture Notes in Mathematics 523. Berlin, Heidelberg, New York-Springer-Verlag.

3. Asorey, M. & Mitter, P.K. (1981). *Regularized, continuum Yang-Mills process and Feynman-Kac functional integral.* Commun. Math. Phys. 80, pp.43-58.

4. Cooper, F. & Freedman, B. (1983). *Aspects of supersymmetric quantum mechanics.* Annals of Phys., 146, pp.262-288.

5. De Angelis, G.F., Jona-Lavinio, G., Sirugue M. (1983). *Probabilistic solution of Pauli type equations.* J. Phys. A. Math. Gen. 16, pp.2433-2444.

6. DeWitt, B.S. *Supermanifolds.* Cambridge University Press, in press.

7. DeWitt-Morette, C. & Elworthy, K.D. (1979). *Stochastic differential equations: lecture notes and problems.* Armadillo Press, Centre for Relativity, University of Texas, Austin, Texas. Revised version in Stochastics ifferential Equations, Materialien XXIII, Schwerpunkt Mathematisierung, Universität Bielefeld, 4800 Bielefeld 1, West Germany.

8. DeWitt Morette, C. & Elworthy, K.D. (1981), editors. *New Stochastic Methods in Physics.* Physics Reports, 77, no. 3.

9. DeWitt-Morette, C., Elworthy, K.D., Nelson, B.L., & Sammelman, G.S.(1980). *A stochastic scheme for constructing solutions of the Schrödinger equation.* Ann. Inst. Henri Poincaré, Section A, 32, no.4, pp.327-341.

10. DeWitt-Morette, C., Maheshwari, A., & Nelson, B. (1979). *Path integration in non-relativistic quantum mechanics.* Physics Reports, 50, no.5, pp.255-372. Amsterdam: North-Holland.

11. Dudley, R.M. (1974). *Recession of some relativistic Markov processes.* Rocky Mountain J. of Math., 4, no.3, pp.401-406.

12. Dudley, R.M., Feldman, J. & LeCam, L. (1971). *On seminorms and probabilities, and abstract Wiener spaces.* Ann. Math. 93, pp.390-408.

13. Elworthy, K.D. (1975). *Measures on infinite dimensional manifolds.* In *Functional Integration and its Applications,* ed. A.M. Arthurs, pp.60-68. Oxford: Oxford University Press.

14. Elworthy, K.D. (1981). *Stochastic methods and differential geometry.* In Seminaire Bourbaki vol. 1980/81. Exposés 561-578, pp.95-110. Lecture Notes in Maths., 901, Berlin, Heidelberg, New York: Springer-Verlag.

15. Elworthy, K.D. (1982). *Stochastic Differential Equations on Manifolds.* *London Math. Soc.* Lecture notes in Mathematics. Cambridge.

16. Elworthy, K.D. & Truman, A. (1981). *Classical mechanics, the diffusion (heat) equation and the Schrödinger equation on a Riemannian manifold.* J. Math. Phys. 22, no. 10, pp.2144-2166.

17. Elworthy, K.D. & Truman, A. (1982). *A Cameron-Martin formula for Feynman Integrals (the origin of the Maslov indices).* In *Mathematical Problems in Theoretical Physics.* Proc. of VIth Int. Conf. on Mathematical Physics, Berlin (West) August, 1981, ed. R. Schrader et al. pp.288-264. Lecture Notes in Physics 153. Berlin, Heidelberg, New-York: Springer-Verlag.

18. Elworthy, K.D. & Truman, A. (1983). *Feynman maps, Cameron-Martin formulae and anharmonic oscillators.* Preprint.

19. Feynman, R.P. & Hibbs, A.R. (1965). *Quantum Mechanics and Path Integrals.* New York: McGraw-Hill.

20. Fujiwara, D. (1980). *Remarks on convergence of the Feynman path integrals.* Duke Math. J., 47, no.3, pp.559-600.

21. Gaveau, B. & Trauber, P. (1981). *Une construction de la quantification euclidienne du Champ de Yang-Mills régularisé.* J. Funct. Anal. 42, pp.356-367.

22. Goldberg, S.I. (1962). *Curvature and Homology.* New York, London: Academic Press.

23. Gowrisankaran, C. (1972). *Quasi-invariant Radon measures on groups.* Proc. Amer. Math. Soc., 35, no.2, pp.503-506.

24. Klein, A., & Perez J.F. (1983). *Supersymmetry and dimensional reduction: a non-perturbative proof.* Phys. Letters B. 125, no.6, pp.473-475.

25. Kobayashi, S. (1957). *Theory of connections.* Annali di Mat., 43, pp.119-194.

26. Kuo, H.-H. (1975). *Gaussian measures in Banach spaces.* Lecture Notes in Maths. 463. Berlin, Heidelberg, New York: Springer-Verlag.

27. Reed, M. & Simon, B. (1972). *Methods of Modern Mathematical Physics I: Functional Analysis.* New York, San Francisco, London: Academic Press.

28. Schwartz, L. (1971). *Probabilités cylindriques et applications radonifiantes.* J. Fac. Sci. Univ. Tokyo Sect. 1, 18, no.2, pp.139-286.

29. Segal, I.E. (1965). *Algebraic Integration Theory.* Bull. Am. Math. Soc., 71, pp.419-489.

30. Simon, B. (1979). *Functional Integration and Quantum Physics.* London, New York: Academic Press.

31. Tarski, J. (1981). *Path integrals over phase space, their definition and simple properties.* Trieste preprint 1C/81/192 (to be published in the proceedings of the summer school in Poiana-Brasov, 1981). International Centre for Theoretical Physics, Miramare, Trieste, Italy.

GRADED MANIFOLDS AND SUPERMANIFOLDS

Majorie Batchelor

Dep. of Pure Mathematics and Mathematical Statistics, University of Cambridge

PREFACE AND CAUTIONARY NOTE

In setting out to write this summary I have tried to tell everything I knew about supermanifolds. I discovered that some of the theorems I had intended to include were without proof, and still others had sketchy proofs with many details left unverified. My policy has been to include some of these results under the heading of conjectures, indicating sketch proofs where they exist. Results called theorems, propositions, corollaries or lemmas are either given with proofs, have proofs in the literature, or I have at some point verified the details. As always, it is a good idea to check the proof of any result before using it.

The enclosed table is intended as an index to the text, to allow readers easy access to the facts they need.

STANDARD NOTATION

All work is done over the real numbers R, but can be generalized to the complex numbers C. (Caution: the generalization of smooth real-valued functions to the complex case is the smooth complex-valued functions, NOT the holomorphic functions.)

0.1 Grading.

If V is a Z_2 (or Z or $Z\times Z_2$ or $Z_2\times Z_2$) graded vector space then

$$V = \bigoplus_{i\in Z_2} V_i \qquad (i \text{ is in } Z,\ Z\times Z_2,\ \text{or } Z_2\times Z_2)\ .$$

H.-J. Seifert et al. (eds.), Mathematical Aspects of Superspace, 91–133.

Say $|v| = i$ if v is in V_i .

If V, W are graded vector spaces so is

$$V \oplus W \qquad (V \oplus W)_i = V_i \oplus W_i$$

$$V \otimes W \qquad (V \otimes W)_i = \sum_{j+k=i} V_j \otimes W_k$$

The twist map

$$T: V \otimes W \to W \otimes V$$

sends

$$v \otimes w \to (-1)^{|v,w|} w \otimes v$$

where if V, W are Z or Z_2 graded then

$$|v,w| = |v||w|.$$

It is of course possible that v is not an element of V_i for any i. To be correct one should always assume that v is in $V_{|v|}$. This assumption will always be made.

If V, W are bi-graded there are two popular conventions. If

$$|v| = (i,j)$$

$$|w| = (h,k)$$

then either

$$|v,w| = ih + jk \qquad \text{CONVENTION A}$$

or

$$|v,w| = ih + ik + jh + jk \qquad \text{CONVENTION B.}$$

Happily most identities can be checked using $|v,w|$ formally, and thus hold in either convention.

0.2 Graded Commutative Algebras.

An algebra A is graded if it is graded as a vector space and satisfies

$$|ab| = |a| + |b| \text{ for } a,b \text{ in } A.$$

If in addition

$$ab = (-1)^{|a,b|} ba$$

A is said to be graded commutative. Algebra homomorphisms are always assumed to preserve the grading.

0.3 <u>Manifold</u>.

"Nice manifolds" are smooth (C^∞) paracompact Hausdorff manifolds. All manifolds are assumed to be nice unless otherwise indicated. If X is a nice manifold, $C^\infty(X)$ denotes the smooth functions on X.

0.4 <u>Sheaves</u>.

A sheaf S over a topological space X is a machine which associates to every open set U of X a linear space (or algebra, or ring or module or other structure) S(U) with restriction maps

$$\rho(U,V): S(U) \to S(V) \quad \text{for open sets } U \supset V \text{ in } X$$

which satisfy the following properties.

I. If $U \supset V \supset W$ are open sets then
$$\rho(V,W)\rho(U,V) = \rho(U,W) .$$

II. If $U = \bigcup_\alpha U_\alpha$ with U_α open sets in X and if s_α is in $S(U_\alpha)$ such that
$$\rho(U_\alpha,U_\alpha \cap U_\beta)s_\alpha = \rho(U_\beta,U_\alpha \cap U_\beta)s_\beta$$
then there exists s in S(U) with
$$\rho(U,U_\alpha)s = s_\alpha$$
for all α.

III. With U, and U_α as in II, if s and s' are in S(U) with
$$\rho(U,U_\alpha)s = \rho(U,U_\alpha)s'$$
for all α, then $s = s'$.

The example to keep in mind is the sheaf of smooth functions over a nice manifold.

A map of sheaves over X

$$\sigma: S \to T$$

is a collection of maps (preserving appropriate structure)

$$\sigma(U): S(U) \to T(U)$$

for all open sets U such that the maps $\sigma(U)$ commute with the restriction maps. Say σ is surjective (injective) if for every point x in X there is a neighbourhood U of x such that $\sigma(U)$ is surjective (injective).

If S is a sheaf on X, and U is an open set in X, S restricts to a sheaf $S|_U$ on U in the obvious manner.

0.5 Vector Spaces.

Spanning.

The subspace of a vector space V spanned by elements $v_1,\dots,v_n$ will often be denoted by brackets $\langle v_1,\dots,v_n\rangle$.

Dual Spaces.

The full linear dual of a vector space V will be denoted by V'. Brackets will often be used to denote evaluation, thus for v in V and α in V', write $\langle\alpha,v\rangle$. When the vector spaces are modules over graded algebras care must be taken with the ordering, thus:

$$\langle\alpha,v\rangle = (-1)^{|\alpha,v|}\langle v,\alpha\rangle$$

0.6 Exterior algebras.

If V is a vector space. Let ΛV denote the exterior algebra on V, the Z graded commutative algebra freely generated by elements of V with $V < (\Lambda V)_1$. If v_1, v_2, ... form a basis for V then set

$$(\Lambda V)_k = \{v_{i_1}\wedge\dots\wedge v_{i_k} : 1 \leq i_1 < i_2 < \dots\ i_k\}.$$

Then $\Lambda V = \bigoplus_k (\Lambda V)_k$, and a basis for V yields a basis for ΛV. As always, the Z grading on ΛV induces a Z_2 grading on ΛV.

Section 1. The category GM.

1.1 Definitions and examples of graded manifolds.

The idea is borrowed from algebraic geometry, that most of geometry can be done given an adequate knowledge of the function ring. In the light of this principle graded manifolds can be defined as follows.

1.1.1 Definition.

A graded manifold is a pair (X,A) where X is a nice manifold and A is a sheaf of Z_2 graded commutative algebras such that

i) There is a surjective map of sheaves

$$\varepsilon: A \to C^\infty.$$

ii) There is an open cover (U_α) of X and isomorphisms of sheaves

$$\tau_\alpha : A|_{U\alpha} \to C^\infty \otimes \Lambda R^s|_{U\alpha}$$

where ΛR^s is the exterior algebra on s generators. If X is an r-dimensional manifold, (X,A) is a graded manifold of dimension (r,s), and the odd dimension of X is s.

1.1.2 Examples.

i) (X,C^∞) where $C^\infty(U)_0 = C^\infty(U)$ for any open set U in X.

ii) $(*, R^s)$ where * is the space consisting of a single point.

iii) (X,Ω) where Ω is the sheaf of smooth differential forms on a nice manifold X.

iv) If E is a vector bundle over X, form the exterior vector bundle ΛE. Then if $\Gamma(\ ,\Lambda E)$ is the sheaf of smooth sections of ΛE, $(X, \Gamma(\ ,\Lambda E))$ is a graded manifold.

1.1.3 Theorem.

If (X,A) is a graded manifold there is a vector bundle E over X such that A is isomorphic to $\Gamma(\ ,\Lambda E)$ as sheaves of Z_2 graded commutative algebras.

Proof. See Batchelor [1]. Roughly speaking the proof is as follows. If (X,A) is a graded manifold, by condition ii there is a cover $\{U_\alpha\}$ and isomorphisms

$$\tau_\alpha : A|_{U_\alpha} \to C^\infty \otimes \Lambda R^s|_{U_\alpha}.$$

Thus $\tau_\alpha\tau_\beta^{-1} = \rho_{\alpha\beta}$ provides an automorphism of $C^\infty(U_\alpha \cap U_\beta) \otimes \Lambda R^s$, and $\rho_{\alpha\beta}\rho_{\beta\gamma}\rho_{\gamma\alpha} = 1$. Such a collection $\{\rho_{\alpha\beta}\}$ represents an element of the Čech cohomology set $H^1(X, \mathrm{Aut}(C^\infty(\ ) \otimes \Lambda R^s))$. It can be verified that this yields a one-one correspondence

$$\{\text{Sheaves A over X satisfying 1.1.2}\} \longleftrightarrow H^1(X, \mathrm{Aut}(C^\infty(\ ) \otimes \Lambda R^s)).$$

The next problem is to calculate $\mathrm{Aut}(C^\infty(U) \otimes R^s)$ for an open neighbourhood U in R^r. This calculation yields a chain of normal subgroups

$$1 \vartriangle K_s \vartriangle K_{s-1} \vartriangle \dots \vartriangle K_1 \vartriangle \mathrm{Aut}(C^\infty(U) \otimes \Lambda R^s)$$

where

$$K_j = \{\sigma \in \mathrm{Aut}(C^\infty(U) \otimes \Lambda R^s) :$$

σ induces the identity on

$$C^\infty(U) \otimes \Lambda R^s / \{ \bigoplus_{p<j} C^\infty(U) \otimes \Lambda^p R^s \}$$

(Recall that a normal subgroup H of a group G is a subgroup such that if h is in H, so is $g^{-1}hg$ for all g in G. Thus normal subgroups are exactly those for which the set of cosets itself forms a group.)

Using the fact that the sheaves K_j/K_{j+1} are fine for $j = 1,\ldots,s$, and $\mathrm{Aut}(C^\infty(U) \otimes \Lambda R^s)/K_1$ is isomorphic to the sheaf GL(U,s) which assigns to U the group of smooth functions from U into the general linear group $GL(R^s)$, it is then shown that there are isomorphism

$$H^1(X, \mathrm{Aut}(C^\infty(\ ) \otimes \Lambda R^s)) \cong H^1(X, \mathrm{Aut}(C^\infty(\ ) \otimes \Lambda R^s)/K_1)$$
$$\cong H^1(X, GL(\ ,s)).$$

This last set is known to correspond to vector bundles of dimension s over X. See also 1.2.9.

1.1.4 Definitions.

A map σ: (X,A) → (Y,B) of graded manifolds is an algebra homomorphism σ: B(Y) → A(X). The category of graded manifolds and their maps is denoted by GM.

The appropriate analogue of point set in the category GM is given by the dual coalgebra.

1.1.5 Definition Proposition.

If A is an algebra over R define

$A^o = \{\alpha: A \to R$: α is linear, there exists an ideal J in A, such that ker $\alpha > J$, and A/J is finite dimensional$\}$.

Then A^o has the following properties.

i) If A, B are algebras, $A^o \otimes B^o = (A \otimes B)^o$.

ii) If f: A → B is an algebra homomorphism then f induces a map

$$f^o: B^o \to A^o .$$

iii) Multiplication μ: $A \otimes A \to A$ induces a comultiplication

$$\Delta: A^o \to A^o \otimes A^o$$

giving A^o the structure of a coalgebra.

Thus for a^o in A, $\Delta a^o = \sum_i a^o_{1i} \otimes a^o_{2i}$ is the unique element of $A^o \otimes A^o$ satisfying

$$a^o(ab) = \sum_i a^o_{1i}(a) a^o_{2i}(b)$$

for all a, b in A.

The map f^o of ii is a map of coalgebras.

The coalgebra A^o is called the dual coalgebra of A. For proofs and a further description of the dual coalgebra see Sweedler [14].

1.1.6 Examples-Proposition.

i) Let U be an open set in R^r. If $\mu = (i_1,\ldots,i_r)$ is an r-tuple of non-negative integers write

$$\partial(\mu)(p) = \left.\frac{\partial^{\Sigma i_j}}{\partial x_1^{i_1}\cdots\partial x_r^{i_r}}\right|_p$$

Then $C^\infty(U)^o$ is the linear space spanned by elements $\partial(\mu)(p)$ for p in U. The comultiplication in $C^\infty(U)^o$ is given by the general Leibniz rule

$$\Delta(\partial(\mu)(p)) = \Sigma\partial(\mu')(p) \otimes \partial(\mu'')(p)$$

over all $\mu' = (i_1',\ldots,i_r')$, $\mu'' = (i_1'',\ldots,i_r'')$ with $i_j' + i_j'' = i_j$.

ii) If L is a finite dimensional vector space and L' is its linear dual, with basis $\lambda_1,\ldots,\lambda_s$, then

$$(\Lambda L)^o = \Lambda L'$$

and

$$\Delta\lambda_1^{h_1}\wedge\ldots\wedge\lambda_s^{h_s} = \Sigma\sigma(\underline{i},\underline{j})\lambda_1^{i_1}\wedge\ldots\wedge\lambda_s^{i_s} \otimes \lambda_1^{j_1}\wedge\ldots\wedge\lambda_s^{j_s}$$

where i_k, j_k, h_k are all in $\{0,1\}$, $i_k + j_k = h_k$, and $\sigma(\underline{i},\underline{j})$ is the sign defined by the equation

$$\lambda_1^{h_1}\wedge\ldots\wedge\lambda_s^{h_s} = \sigma(\underline{i},\underline{j})\lambda_1^{i_1}\wedge\ldots\wedge\lambda_s^{i_s}\wedge\lambda_1^{j_1}\wedge\ldots\wedge\lambda_s^{j_s}.$$

iii) If (X,A) is a graded manifold and U is an open set in X, then $A(U)^o$ separates elements of A(U); that is, the map

$$A(U) \to (A(U)^{o})'$$

is injective.

Proof.

i) The inclusion $j: P \to C^{\infty}(U)$ of the polynomials into the smooth functions on U gives a map

$$j^{o}: C^{\infty}(U)^{o} \to P^{o} .$$

The coalgebra P^{o} is very easy to calculate as it is freely generated by commuting elements $x_1,\ldots,x_r$. The result is that P^{o} is the linear space spanned by the elements $\partial(\mu)(p)$ for p in R^{r}. The next step is to verify that the image of j^{o} is the linear space spanned by elements $\partial(\mu)(p)$ with p in U. The last and more difficult step is to show that j^{o} is injective.

ii) Since L is finite dimensional so is ΛL and $(\Lambda L)' = \Lambda L' = (\Lambda L)^{o}$.

iii) See Kostant [9].

1.2 Bundles in GM.

The motivation for the definition of bundle is the characterization of ordinary smooth vector bundles over a nice manifold as locally free modules over the sheaf of smooth functions. The idea is that just as one can regard the sheaf of functions as the primary object rather than the point set of a manifold, so one can regard the sheaf of sections of a smooth vector bundle as the primary object rather than the vector bundle itself.

1.2.1 Definition.

Let (X,A) be a graded manifold. A vector bundle over (X,A) is a sheaf M of locally free modules over A. That is, for any point p in X there is an open neighbourhood U of p such that

$$M(U) = A(U)m_1 \oplus \cdots \oplus A(U)m_n$$

for some $m_1,\ldots,m_n$ in M(U). (For more information on locally free sheaves and their correspondence with vector bundles see Wells [15].)

1.2.2 Examples.

i) The tangent bundle. Set

$$\mathrm{Der}(U,A) = \{\alpha : A(U) \to A(U) : \alpha \text{ is R-linear}, \quad \alpha(fg) = \alpha(f)g \oplus (-1)^{|\alpha||f|} f\alpha(g)\} .$$

Use the assignment to build a sheaf. First define restriction maps

$$\tau(U,V): \mathrm{Der}(U,A) \to \mathrm{Der}(V,A)$$

for open sets $V < U$. If $\rho(U,V): A(U) \to A(V)$ is the restriction map, notice that $\rho(U,V)$ is onto. For b in A(V) and γ in Der(U,A) define

$$\tau(U,V)\gamma(b) = \rho(U,V)\gamma(a)$$

where a is in A(U) and $\rho(U,V)a = b$. Well definition depends on the fact if $\rho(U,V)a = \rho(U,V)a'$ then $\rho(U,V)\gamma a = \rho(U,V)\gamma a'$. For more information on the structure of A(U) and Der(U,A) see Kostant [9].

So far Der(,A) has the structure of a presheaf; that is, it has restriction maps which can be shown to have property i for sheaves. Now use a standard construction which constructs a sheaf from a given pre-sheaf. See Wells [15].

ii) The bundle of forms on (X,A). Define $\underline{\Omega}(U,A)$ to be the $Z \times Z_2$ bi-graded commutative algebra with

$$\underline{\Omega}(U,A)_{(o,*)} = A(U)$$

$$\underline{\Omega}(U,A)_{(1,j)} = \{\Sigma a_i db_i : a_i, b_i \in A(U),\ |a_i| + |b_i| = j\}$$

subject to the condition

$$d(ab) = (da)b + (-1)^{|d,a|} a(db).$$

That is, d is to be a derivation of degree (1,0).

Since A is a sheaf, $\underline{\Omega}(,A)$ is a presheaf. Let $\Omega(,A)$ be the sheaf generated by $\underline{\Omega}(,A)$. Notice that $\Omega(,A)$ has all the formal properties expected of forms. The following proposition identifies $\Omega(U,A)_{(1,*)}$ as the dual bundle to the tangent bundle.

1.2.3 Proposition.

There is an A(U) linear map

$$k_{(b,*)}: (U,A)_{(b,*)} \to \mathrm{Hom}_{A(U)}(\otimes^b_{A(U)} \mathrm{Der}(U,A),\ A(U))$$

mapping $\Omega(U,A)_{(b,*)}$ isomorphically onto the A(U) linear maps α satisfying

$$(\xi_1 \otimes \dots \otimes \xi_i \otimes \xi_{i+1} \otimes \dots \otimes \xi_b,\ \alpha) = (-1)^{\sigma}(\xi_1 \otimes \dots \otimes \xi_{i+1} \otimes \xi_i \otimes \dots \otimes \xi_b,\ \alpha)$$

where $\sigma = 1 + |\xi_i,\xi_{i+1}|$.

In particular,

$$k_{(1,*)}: (U,A)_{(1,*)} \to \mathrm{Hom}_{A(U)}(\mathrm{Der}(U,A),A(U))$$

identifies $\Omega(U,A)_{(1,*)}$ with the dual bundle to the tangent bundle.

Proof.

See Kostant's section on forms and Batchelor [3]. The idea is that $\Omega(U,A)$ has universal properties that guarantee the existence of the maps $k_{(b,*)}$. To see that the maps are isomorphisms it is simplest to compute both spaces in the case that U is an open set with $A(U) \cong C^{\infty}(U) \otimes \Lambda R^s$.

Vector bundles over (X,A) always give rise to vector bundles over X in the usual sense.

1.2.4 Proposition.

Let M be a vector bundle over (X,A). Then there is a vector bundle ξ:E → X such that if $\Gamma(\ ,E)$ is the sheaf of smooth sections of E there is a map of sheaves (of A modules)

$$r: M \to \Gamma(\ ,E).$$

Proof.

Define NA to be the sheaf of A modules which assigns to U the ideal of nilpotent elements of A(U). Define

$$\underline{M}(U) = M(U)/NA(U)M(U).$$

Now $\underline{M}$ is a sheaf of $A/NA \cong C^\infty$ modules. $\underline{M}$ is locally free as a C^∞ module since M is free as an A module. Locally free C^∞ modules correspond to vector bundles: let E denote the vector bundle associated with $\underline{M}$. The A module structure on E is given by the homomorphism $\varepsilon: A \to C^\infty$. The map r is simply the projection map $M \to M/(NA)M$.

1.2.5 Example-Proposition.

The tangent bundle. Define

$$T_p(X,A) = \{\alpha:A(X) \to R:\alpha(ab) = \alpha(a)\varepsilon b(p) + (-1)^{|a,\alpha|}\varepsilon a(p)\alpha(b)\}$$

where ε is the map $\varepsilon:A(X) \to C^\infty(X)$ given in the definition of graded manifolds.

$$T(U,A) = \cup\, T_p(X,A) \text{ for } p \text{ in an open set } U \text{ in } X.$$

So far $T(X,A)$ is a union of vector spaces. It can be given a smooth structure by defining the "smooth" sections of $T(X,A)$. A map

$$s:U \to T(U,A)$$

satisfying

$$s(p) \in T_p(X,A)$$

is smooth if for every a in $A(U)$ the map $s_a: U \to R$ given by

$$s_a(p) = \langle s(p),a\rangle$$

is smooth.

Proposition.

The real vector bundle over X associated with the tangent bundle over (X,A) is $T(X,A)$.

Proof.

There is a map of sheaves

$$r: \mathrm{Der}(\ ,A) \to \Gamma(\ ,T(X,A))$$

given by

$$r\gamma(p)(a) = \varepsilon\gamma(a)(p)$$

for γ in Der(U,A). Thus NA(U)Der(U,A) is in the kernel of r(U). To see that $\Gamma($,T(X,A)) is isomorphic to Der(,A)/NADer(,A) check that for a neighbourhood U in X with Der(U,A) a free A(U) module, a free set of A(U) generators of Der(U,A)maps isomorphically onto a free set of C^∞ generators of Γ(U,T(X,A)).

1.2.6 Example-Proposition.

Define

T(X,A)' = the dual bundle of T(X,A)

ΛT(X,A)' = the exterior bundle of T(X,A)'.

Then if Ω_A denotes the sheaf of smooth sections of ΛT(X,A)' then

$$\Omega_A \cong \Omega(\ , \ )/NA\Omega(\ ,A).$$

Proof.

See Kostant's remarks on Ω(,A) and Ω_A.

1.2.7 Remark.

The construction 1.2.4 is useful in that one can now talk about elements in a vector bundle rather than just the sections of a vector bundle. In particular, think of elements in T(X,A) as elements of the tangent bundle.

1.2.8 Proposition.

Let E be a vector bundle over X. A choice of homomorphism

$$\lambda : C^\infty \to A$$

satisfying $\varepsilon\lambda = 1$ on C^∞ allows one to construct a vector bundle

$$M_E = A \otimes_{C^\infty} \Gamma(\ ,E)$$

over (X,A).

If, given a vector bundle M over (X,A), the vector bundle over X described in 1.2.4 is denoted by E_M, then there is an isomorphism

$$M \cong M_{E_M}$$

and a canonical isomorphism

$$E \cong E_{M_E}.$$

1.2.9 Theorem.

(Batchelor [4].) Let (X,A) be a graded manifold and let E be the vector bundle dual to the vector bundle $T(X,A)_1$. Then A is isomorphic to the sheaf $\Gamma(\ ,E)$.

Proof.

This will follow from three lemmas.

Lemma A.

For all p in X there is a subcoalgebra G_p of $A(X)^o$ with G_p isomorphic to $\Lambda T(X,A)_1$. Moreover $G = \bigcup_p G_p$ has a natural vector bundle structure and G is isomormphic to $\Lambda T(X,A)_1$ as vector bundles.

Lemma B.

Let F be a (finite dimensional) vector bundle and let F' be its dual vector bundle. Let $\Gamma(\ ,F)$, $\Gamma(\ ,F')$ be the C^∞ modules of smooth sections of F, F' respectively. Then

$$\Gamma(X,F') = \{\alpha : \Gamma(\ ,F) \to C^\infty(\ ) : \alpha \text{ is } C^\infty \text{ linear}\}.$$

Lemma C.

The inclusion $A(X) \to (A(X)^o)'$ gives an isomorphism

$$\Phi : A(X) \to \{\alpha : \Gamma(\ ,G) \to C^\infty : \alpha \text{ is } C^\infty \text{ linear}\}.$$

To prove lemma A. If p is in X let $A^o(p)$ be the maximal subcoalgebra containing $\mathbb{R}p$ as its unique simple subcoalgebra. (Thus $A^o(p)$ is the space of all multiple differentiations at p.) Let Δ be the comultiplication and set

$$G(0,p) = \mathbb{R}p$$
$$G(k,p) = \{\alpha \in A^o(p) : \text{the component of } \Delta^{k-1}\alpha \text{ in} \otimes^k T_p(X,A)_1 \text{ is non zero, unless } \alpha = 0\}.$$
$$G(k) = \bigcup_p G(k,p)$$
$$G(p) = \bigoplus_k G(k,p)$$
$$G = \bigcup_p G(p) = \bigoplus_k G(k)$$

To see that $G(p)$ is isomorphic to $\Lambda T_p(X,A)_1$ choose a neighbourhood U of p with $A(U)$ isomorphic to $C^\infty(U) \otimes \Lambda\mathbb{R}^s$

and calculate.

The vector bundle structure on G is given by defining the smooth sections $\Gamma(U,G)$ of G over an open set U. Say σ is smooth if the map

$$\sigma_a : U \to R$$

given by

$$\sigma_a(u) = \langle \sigma(u), a \rangle$$

is smooth for any element a in A(U).

To prove Lemma B.

The only tricky part is to show that if $\alpha : \Gamma(\ ,F) \to C^\infty$ is C^∞ linear then α defines an element in $\Gamma(X,F')$. Do this as follows. If s_p is in F_p, define $\underline{\alpha}_p$ in F'_p by setting

$$\langle \underline{\alpha}_p, s_p \rangle = \alpha s(p)$$

for some section s of F over a neighbourhood U of p. Check that $\underline{\alpha}_p$ is independent of the choice of section.

To do this, let $s_1,\ldots,s_r$, $t_1,\ldots,t_r$ be sections of F over U such that the s_i and the t_i form a C^∞ free generating set for $\Gamma(U,F)$ with $s_i(p) = t_i(p)$. Then write

$$s_i = \Sigma f_{ij} t_j$$

and notice that

$$\alpha s_i(p) = \Sigma f_{ij} \alpha t_j(p)$$

$$= \Sigma f_{ij}(p)\ \alpha t_j(p)$$

$$= \alpha t_i(p)$$

since $s_i(p) = \Sigma f_{ij}(p) t_j(p)$ implies that $f_{ij}(p) = 0$ if $i \neq j$ and $f_{ii} = 1$.

To prove Lemma C.

If a is in A(X) and g is in $\Gamma(U,G)$, define

$$\Phi a(g)(u) = \langle a, g(u) \rangle.$$

To check that this map is an isomorphism by showing that the map

$$\Phi|_U: A(U) \to \{\alpha:\Gamma(\ ,G)|_U \to C^\infty|_U: \alpha \text{ is } C^\infty|_U \text{ linear}\}$$

is an isomorphism on neigbourhoods U with $A(U) \cong C^\infty(U) \otimes \Lambda R^s$. This identifies A(U) with $\Gamma(\ ,\Lambda E)$. To see that the map is a map of algebras, notice that the multiplication in $\Gamma(\ ,\Lambda E)$ is the multiplication induced by the comultiplication in G.

Section 2. The geometric approach.

2.1 The general idea.

By contrast to the algebraic approach described in section 1, here supermanifolds are defined as topological spaces which are locally homeomorphic to an analogue of Euclidean space. A category of supermanifolds can be defined by giving the following information:

1. Supereuclidean space $E^{r,s}$ for non-negative integers r, s.
2. A topology on $E^{r,s}$.
3. A smoothness criterion. That is, if U is an open set in $E^{r,s}$ and if V is an open set in $E^{r,s}$, one needs a way of recognizing a subset S(U,V) of the set of all functions from U to V which is to be considered the set of smooth functions.

2.1.1 Definition.

Let S denote a choice of a definition of supereuclidean space together with a topology and a smoothness criterion. Define an S supermanifold to be a topological space X together with an atlas $\{U_\alpha,\varphi_\alpha\}$ of S-smooth charts. That is $\{U_\alpha\}$ is an open cover of X and φ_α is a homeomorphism from U_α to an open set in some supereuclidean space $E^{r,s}$ such that

$$\varphi_\alpha\varphi_\beta^{-1}:\varphi_\beta(U_\alpha \cap U_\beta) \to \varphi_\alpha(U_\alpha \cap U_\beta)$$

is S-smooth.

The problem is thus to provide suitable definitions of supereuclidean space together with topologies and smoothness criteria on it. At least five categories of supermanifold are available. SSM, SM, G^∞, G^ω, and JP, each differing only with respect to these constructions.

2.2 The graded commutative algebra B and supereuclidan space.

The principle here is that in order to end up with graded commutative functions, one must replace the real or complex numbers by a suitable graded commutative algebra B.

2.2.1 Examples.

i) $B = \Lambda R^L$ for $L < \infty$. Thus B is generated by L generators $\gamma_1, \ldots, \gamma_L$. There is an obvious R basis for B with

$$B_o = \langle \gamma_{i_1} \wedge \ldots \wedge \gamma_{i_k} : 1 \leq i_1 < \ldots < i_k \leq L.\ k \text{ even} \rangle$$

$$B_1 = \langle \gamma_{i_1} \wedge .\ . \wedge \gamma_{i_k} : 1 \leq i_1 < \ldots < i_k \leq L.\ k \text{ odd} \rangle.$$

Multiplication in R^L is given by justaposition:

$$(\gamma_{i_1} \wedge \ldots \wedge \gamma_{i_k})(\gamma_{i'_1} \wedge \ldots \wedge \gamma_{i'_{k'}})$$

$$= \gamma_{i_1} \wedge \ldots \wedge \gamma_{i_k} \wedge \gamma_{i'_1} \wedge \ldots \wedge \gamma_{i'_{k'}} .$$

It is not hard to check that the product of two basis elements is either 0 (if some $i'_s = i_r$) or can be arranged (using the identity $r_m \wedge r_n = - r_n \wedge r_m$) to be plus or minus a basis element.

One can define a symmetric bilinear form on B by declaring elements of the above basis to be orthonormal. This gives B the structure of a Banach algebra. Let B^L denote this algebra.

ii) Let V be an infinite dimensional vector space with a countable basis $\{\gamma_i\}$. If μ is a sequence $(i_1, \ldots, i_k)$ with $1 \leq i_1 < i_2 < \ldots < i_k$ set $\gamma(\mu) = \gamma_{i_1} \wedge \ldots \wedge \gamma_{i_k}$. Thus the set $\{\gamma(\mu)\}$ is a basis for ΛV.

Define

$$B^\infty = \{\Sigma a(\mu)\gamma(\mu) : a(\mu) \in R,\ \Sigma|a(\mu)| < \infty\}.$$

Then B^∞ is a Z_2 graded commutative Banach algebra with a norm $\| \ \|$ given by

$$\|\Sigma a(\mu)\gamma(\mu)\| = \Sigma|a(\mu)|.$$

This is an example of a Z_2 graded commutative algebra which is not a Z graded commutative algebra. See Rogers [11].

2.2.2 The Augmentation map ε.

For any of the graded commutative algebras B defined above there is a unique algebra homomorphism

$$\varepsilon: B \to R$$

given by sending the odd generators to zero. For all the Banach space structures given this is a continuous open mapping.

2.2.3 Topologies on B.

In the examples of 2.2.1 B has a topology given by the Banach space structure on B. However there is another topology, the coarse topology, which is occasionally used. This topology is defined by carrying back the topology on R via the augmentation map. Say a set $U \subset B$ is in the coarse topology on B if $U = \varepsilon^{-1}(V)$ for some open set V in R. This topology is not Hausdorff: two points u,v in B are separated by open sets if and only if u-v is invertible in B. Notice that ε is continuous and open with respect to this topology as well.

Unless otherwise indicated, assume that B is equipped with the Banach space topology.

2.2.4 Supereuclidean space.

Given a Z_2 graded commutative algebra B define (r,s) dimensional supereuclidean space

$$E_B^{r,s} = (B_o)^r \times (B_1)^s.$$

If B has a topology, B_o and B_1 have topologies as subspaces of B, and $E_B^{r,s}$ has the topology of a product of topological spaces. In the examples of 2.2.1 where B has a Banach space structure. $E_B^{r,s}$ also has a Banach space structure.

Using the coarse topology for $B = \Lambda R^L$, this means that a set $U \subset E_B^{r,s}$ is open if and only if $U = \varepsilon^{-1}(V)$ where V is an open set in R^r and

$$\varepsilon: E_B^{r,s} \to R^r$$

is given by $\varepsilon(u_1,\ldots,u_r,v_1,\ldots,v_s) = (\varepsilon(u_1),\ldots,\varepsilon(u_r))$. Generally one writes $E^{r,s}$ for $E_B^{r,s}$.

2.3 Smooth maps on $E^{r,s}$.

The task of this section is to give smoothness criteria for functions from open sets in $E^{r,s}$ to B. These vary with the choice of B and the topology on B.

2.3.1 G^∞ smooth maps. (Rogers)

$B = \Lambda R^L$, $L < \infty$, with the Banach space structure.

Definition.

Let U be an open set in $E^{r,s}$. A continuous map $f: U \to B$ is said to be G^1 if there exist functions

$$H_i, K_j: U \to B, \quad i = 1,\dots,r, \quad j=1,\dots,s$$

such that if (u,v) and $(u+h,v+k)$ are in U, then

$$f(u+h,v+k) = f(u,v) + \Sigma h_i H_i(u,v) + \Sigma k_j K_j(u,v) + o(h,k).$$

The map f is said to be G^p on U if f is G^1 and it is possible to choose H_i, K_j as above with H_i, K_j being G^{p-1}. If f is G^p for all p f is said to be G^∞.

2.3.2 G^ω analytic maps. (Rogers [11])

$B = \Lambda R^L$, $L < \infty$, with the Banach space structure.

Definition.

Let

$$x_i: E^{r,s} \to B$$

be the projections onto the coordinates of $E^{r,s}$. If $\mu = (i_1,\dots,i_r)$ is a r-tuple of non-negative integers write

$$p(\mu)(x) = x_1^{i_1},\dots,x_r^{i_r}$$

A function $g: E^{r,s} \to B$ is represented by the power series $\Sigma a(\mu)p(\mu)$ at p if there is a neighbourhood N of p such that for q in N

$$g(q) = \Sigma a(\mu)p(\mu)(q-p).$$

The power series $\Sigma a(\mu)p(\mu)$ is absolutely convergent on U if the sum

$$\Sigma \|a(\mu)\| \, \|p(\mu)(u)\|$$

is convergent for all u in U.

If U is an open set in $E^{r,s}$, a function g: U → B is G^{ω} if for every p in U there is a neighbourhood N_p of p in U such that g is represented by an absolutely convergent power series at p.

A function f: U → $E^{r,s}$ is G^P, G^{∞}, or G^{ω} if the projections onto each coordinate are G^P, G^{∞}, or G^{ω} respectively.

2.3.3 <u>JP smooth maps</u> (Jadzyck Pilch [8])

B is a Z_2 graded commutative Banach algebra with an augmentation homomorphism

$$\varepsilon: B \to R$$

satisfying the following two conditions.

i) The map sending B_i into $L_{B_o}(B_j, B_{j+i})$, the space of B_o linear maps from B_j to B_{j+1} given by left (right) multiplication by elements of B_i is a norm-preserving isomorphism.

ii) The norm on B satisfies

$$\|b\| = \|b - \varepsilon(b)\| + |\varepsilon(b)|.$$

An example is B = B^{∞} with the Banach space structure given in 2.2.1 ii. See sketch of proof below.

2.3.4 <u>Definition</u>. (Jadzyck Pilch [8])

Let U be an open set in $E^{r,s}$. A map g: U → $E^{r',s'}$ is J^1 if g is continuous and if there are maps

$$G: U \to L_{B_o}(E^{r,s}, E^{r',s'})$$

such that if u and u+h are in U then

$$g(u+h) = g(u) + G(u)h + o(h).$$

By condition i on B, $L_{B_o}(E^{r,s}, E^{r',s'})$ is isomorphic to $E^{rr'+ss', rs'+sr'}$ as B_o modules. Thus we can make the following definition.

2.3.5 Definition.

Let U be an open set in $E^{r,s}$. A map $g: U \to E^{r,s}$ is said to be JP^p if g is continuous and there is a map G, as in 2.3.4 with G of class JP^{p-1}. If g is JP^p for all p, g is said to be JP^∞.

Sketch proof of claim in 2.3.3.

First check that ΛV is a Banach algebra. Now prove condition i for ΛV. Since B_o is generated by the identity a B_o linear map from B_o to B_j is completely determined by the image of 1. Thus $L_{B_o}(B_o,B_j)$ is isomorphic to B_j.

The map $B_j \to L_{B_o}(B_1,B_{j+1})$ given by left multiplication is easily seen to be injective. The problem is to show that it is surjective as well. Let T be in $L_{B_o}(B_1,B_{j+1})$, and let γ be in B_1 with $T\gamma \neq 0$. Then $\gamma T\gamma = 0$: otherwise there would be β in B_1 with

$$0 \neq \beta\gamma T\gamma = T\beta\gamma\gamma = 0.$$

This means that $T\gamma$ can be written as $\alpha\gamma$ for some α in B_j. Let W be the subspace of V consisting of those γ such that $T\gamma = \alpha\gamma$. The aim is to show that W is all of V. Choose β in V with β,γ linearly independent, and suppose that $T\beta = \mu\beta$ and $T(\beta+\gamma) = \nu(\beta+\gamma)$. Comparing $T(\beta+\gamma)$ and $T\beta + T\gamma$ implies

$$\nu\beta + \nu\gamma = \mu\beta + \alpha\gamma.$$

Thus

$$\nu\beta\gamma = \mu\beta\gamma$$

$$\nu\gamma\beta = \alpha\gamma\beta$$

or

$$\alpha - \nu = \rho\gamma\beta$$

$$\nu - \mu = \sigma\beta\gamma$$

for some ρ, σ in B. Thus

$$\alpha = \mu + (\rho-\sigma)\gamma\beta$$

and $\alpha\beta = \mu\beta = T\beta$ as desired. The remaining verifications are easier.

2.3.6 SSM Smooth Maps.

$B = \Lambda R^L$, $L < \infty$, with the coarse topology.

Since B has not got a Banach space structure smooth maps must be defined algebraicly. The idea is to take the smallest subalgebra which contains all maps which are "obviously" smooth. For SSM smooth maps the following types of function are considered smooth.

Let U be a (coarse) open set in $E^{r,s}$ and let F(U) denote the algebra of all functions from U to B. For SSM smooth maps the following subalgebras of F(U) are considered smooth.

i) The algebra P generated over the real numbers by the coordinate projections. Let $J: P \to F(U)$ denote the inclusion.

ii) The algebra $C^\infty(\varepsilon(U))$ included in F(U) in the following way. Let $x_1,\dots,x_r$ be a set of coordinates for R^r. If $\mu = (i_1,\dots,i_r)$ is an r-tuple of non-negative integers write

$$\partial(\mu) = \frac{\partial^{\Sigma i_j}}{\partial x_1^{i_1},\dots, x_r^{i_r}}$$

$$p(\mu)(u_1,\dots,u_r) = u_1^{i_1},\dots,u_r^{i_r}$$

Define $K: C^\infty(\varepsilon(U)) \to F(U)$ by

$$Kf(u_1,\dots,u_r,v_1,\dots,v_s) =$$

$$\Sigma a_\mu \partial(\mu) f(\varepsilon(u_1),\dots,\varepsilon(u_r)) p(\mu)(u_1-\varepsilon(u_1),\dots, u_r-\varepsilon(u_r)) .$$

It should be checked that K is an algebra homomorphism.

Definition.

The set of SSM smooth maps is the image of the map

$$J \otimes K: P \otimes C^\infty(\varepsilon(U)) \to F(U).$$

Notice that if Q is the subalgebra of P generated by the projections onto the odd coordinates. $J \otimes K$ is injective on $Q \otimes C^\infty(\varepsilon(U))$.

2.3.7 SM smooth maps.

$B = \Lambda R^L$, $L < \infty$, with the coarse topology on B.

The problem with SSM smooth maps is that constant B-valued functions are not considered smooth. To repair this, simply add to the list of "approved smooth functions" in 2.3.6 the following set.

iii) The constant functions. Write $C: B \to F(U)$ for the inclusion of B in F(U)

Definition.

The SM smooth functions on a coarse open set $U < E^{r,s}$ are those in the image of the map

$$C \otimes J \otimes K: B \otimes P \otimes C^{\infty}(\varepsilon(U)) \to F(U).$$

As with G^{∞} functions, a function $f: U \to E^{r',s'}$ is SSM (or SM) if f composed with projection onto any coordinate projection is SSM (SM).

2.3.8 Remark.

The definition given in 2.3.6 differs from that given in Batchelor [2]. Both definitions describe the same subalgebra of F(U). While 2.3.6 is not obviously independent of choice of coordinates, it is much the simpler and more intuitively satisfying definition.

2.3.9 Notation.

If U is an appropriate open set in $E^{r,s}$, write $G^{\infty}(U)$, $G^{\omega}(U)$, JP(U), SSM(U), SM(U) for the subset of F(U) consisting of G^{∞}, G^{ω}, JP, SSM, SM, maps respectively. It will be convenient to let S(U) denote smooth maps in whatever category is appropriate, particularly when formally identical constructions hold for several different sets of smooth function. For example, if U is an open set in $E^{r,s}$, write $S(U,E^{r',s'})$ for the appropriate set of smooth maps from U to $E^{r',s'}$.

2.3.10 Proposition.

Let U be an appropriate open set in $E^{r,s}$. Then

i) S(U) is a subalgebra of $F(U)$.

ii) Let V be an open set in $E^{r',s'}$. If f is in S(U,V) and g is in $S(V,E^{r'',s''})$ then gf is in $S(U,E^{r'',s''})$.

Proof.

For S one of G^∞, G^ω, JP, the proofs follow the usual arguments for C^∞ or analytic functions. In the case of SSM maps, see Batchelor [2]. In the case of SM maps regard this proposition as a conjecture.

Finally it is useful to know what the smooth functions "look like".

2.3.11 Proposition.

For B, $E^{r,s}$, and $U \subseteq E^{r,s}$ as appropriate the map

$$C \otimes K \otimes J: B \otimes C^\infty(\varepsilon(U)) \otimes \Lambda R^s \to F(U)$$

maps onto the S smooth functions for S any of G^∞, JP, SM. This map is injective if S is G^∞ with $L = \infty$, or JP. The map

$$K \otimes J: C^\infty(\varepsilon(U)) \otimes \Lambda R^s \to F(U)$$

is an isomorphism onto the SSM smooth maps.

Proof.

In the case of SM or SSM smooth maps this is immediate from the definition. In the case of JP or G^∞ smooth maps see Jadzyck-Pilch [8] or Rogers [11] respectively.

2.4 Examples of supermanifolds.

2.4.1 $E^{r,s}$.

It is useful to have a pictorial method of representing supereuclidean space. I like to draw coordinate axes with solid axes representing commuting coordinates, that is copies of B_0, and dotted axes representing anti-commuting coordinates, or copies of B_1. In the coarse topology open sets are represented as cylinders over open sets in R^r (e.g. X, Y below).

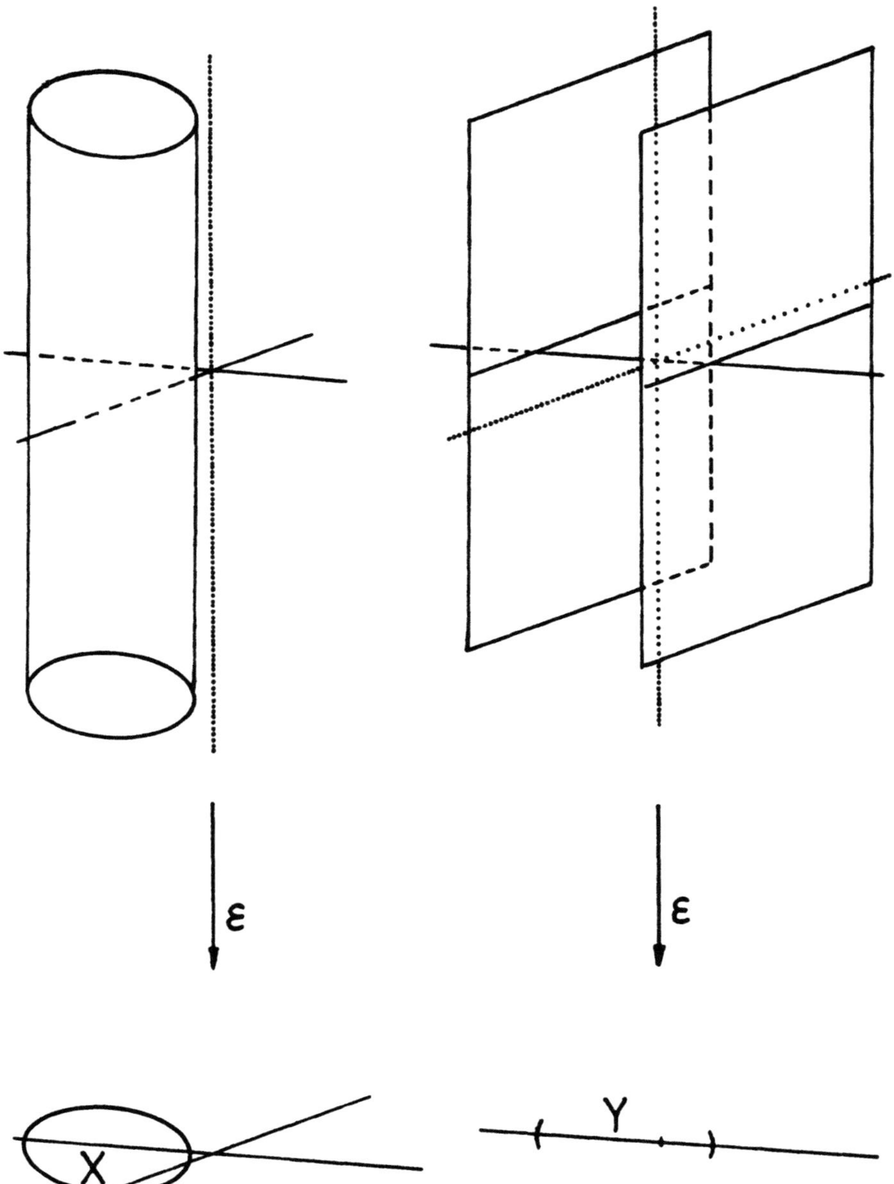
ε
ε
X
Y
$E^{2,1}$
$E^{1,2}$

2.4.2 The Möbius band.

Let $B = \Lambda R^1$. Define $E^{1,1}/\sim$ where $\sim$ is given by $(u,v) \sim (u+n,(-1)^n v)$ for $n \in Z < B$.

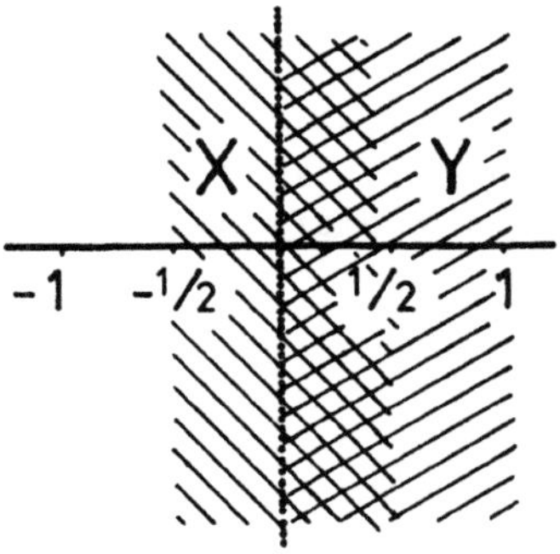

The map $(u,v) \to (u,-v)$ is smooth with respect to any definition of smooth. The open sets X, Y indicated are open with respect to any topology available. Thus M endowed with the fine topology is a G^α or a G^ω supermanifold, and M endowed with the coarse topology is an SM or SSM supermanifold.

2.4.3 The Möbius band again.

Again let $B = \Lambda R^1$, and let v be a chosen odd generator of B. Define $N = E^{1,1}/\sim$, where $\sim$ is given by $(u,w) \sim ((-1)^n u, w+nv)$.

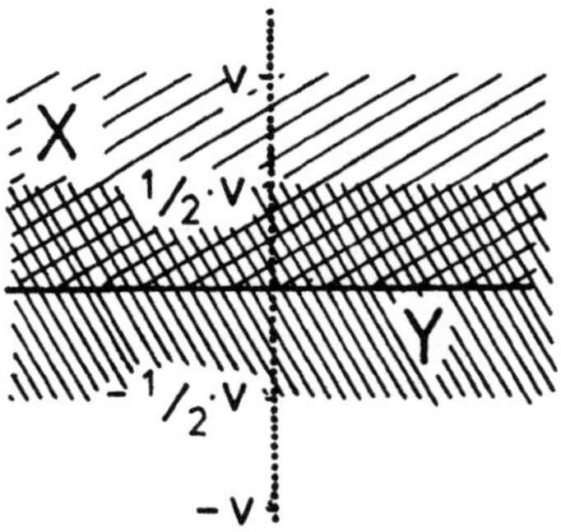

Here the sets X and Y are open in the fine topology but not in the coarse topology. The map $(u,w) \to (-u,w+v)$ is smooth in the G^∞ sense. Thus N is a G^∞ supermanifold.

2.4.4 The Torus.

Again let $B = \Lambda R^1$, and choose an odd generator v.

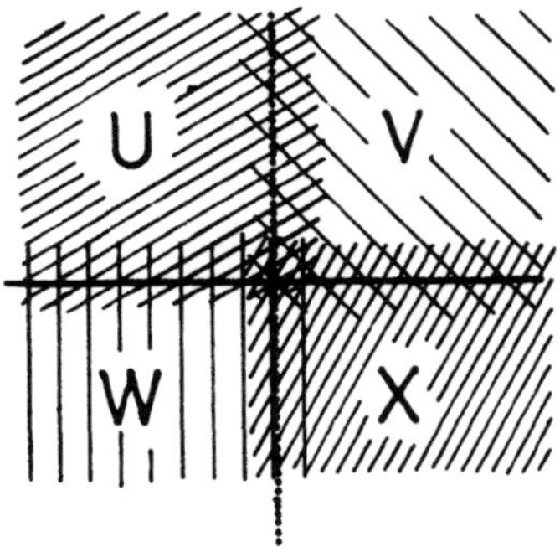

Define $T = E^{1,1}/\sim$, where $\sim$ is given by $(u,w) \sim (u+m,w+nv)$ for m, n in Z. The sets U,V,W,X drawn are open in the fine topology and all patching maps are G^∞.

2.5 Bundles over supermanifolds.

2.5.1 Definition.

Bundles over supermanifolds. Let X, Y be S supermanifolds. An X bundle over Y is an S supermanifold Z with an S map $\pi : Z \to Y$ such that for every point p in Y there is a neighbourhood N of p in Y and an S diffeomorphism

$$\tau : \pi^{-1}(N) \to N \times X$$

with $\tau(x)$ in $(\pi(x),X)$.

2.5.2 Definition.

Vector superspaces. See Jadzyck, Pilch [8]. An (r,s) dimensional vector superspace over B is a B_o module isomorphic to $E_B{}^{r,s}$.

It is convenient for vector superspaces to be considered as even B_o modules. That is, if V is a vector superspace, $V = V_o^o$.

Notice that $E^{r,s}$ can be given a B module structure only if r = s. Also notice that $E^{t,s}$ is not a free B_o module. The next proposition part iii gives a condition which characterized vector superspaces.

2.5.3 Proposition.

i) If V is an (r,s) dimensional vector superspace and if W is an (r',s') dimensional vector superspace, $V \oplus W$ is an (r+r',s+s')dimensional vector superspace.

ii) Suppose M is a free B module on r generators in M_o and s generators in M_1. Then M_o^o is a vector superspace of dimension (r,s) and M_1^o is a vector superspace of dimension (s,r).

iii) A B_o module V is a vector superspace if and only if there are free B modules H, K with H free on even generators and K free on odd generators, and

$$V \cong H_o \otimes K_o.$$

2.5.4 Remark.

The proof of proposition 2.5.3 is not hard. Beware that vector superspaces behave badly under quotients and tensor products. The quotient of a vector superspace by a sub vector superspace is not necessarily a vector superspace. Neither is the tensor product of two vector superspaces a vector superspace.

2.5.5 Definition.

A super vector bundle E over a supermanifold X is an $E^{r,s}$ bundle over X as in 2.5.1 such that the maps

$$\tau|_{\pi^{-1}} : \pi^{-1}(x) \to (x, E^{r,s})$$

are B_o linear isomorphism. Similarly, if M is a free B module, M has the structure of a supermanifold and one can define M bundles over X and insist that the maps

$$\tau|_{\pi^{-1}} : \pi^{-1}(x) \to (x, M)$$

preserve the B module structure on M.

2.5.6 Example.

The tangent bundle for JP (G^∞) supermanifolds. See Jadzyck, Pilch [8]. Let S be a JP supermanifold. Then S can also be considered as a Banach manifold, and the tangent space can be defined in the usual manner as a set of equivalence classes of curves. Define

$$T_s S = \{c(-\varepsilon,\varepsilon) \to S: c(0) = s\}/\sim$$

where $\sim$ is given by

$$c \sim c' \text{ if and only if } \frac{d}{dt}\varphi c|_o = \frac{d}{dt}\varphi c'|_o$$

with respect to some chart (U,φ) with s in U. As usual, the equivalence relation is independent of the choice of chart. Let [c] denote the equivalence class of c.

Now let (U,φ) be a chart with $\varphi(s) = 0$. Use this chart to give $T_s S$ a B_o module structure: Define

$$B[c] = [\varphi^{-1} b\varphi c].$$

Check that this is independent of the chart chosen. If (V,ψ) is another chart with $\psi(s) = 0$ then

$$\frac{d}{dt}\psi\,\varphi^{-1} b\varphi c|_o = D(\psi\varphi^{-1})|_o\, b\,\frac{d}{dt}\varphi c|_o$$

$$\overset{*}{=} b\, D(\psi\varphi^{-1})|_o \frac{d}{dt}\varphi c|_o$$

$$= b \frac{d}{dt} \psi c|_o = \frac{d}{dt} \psi \psi^{-1} b \psi c|_o .$$

Here $D(\psi\varphi^{-1})$ is the ordinary Jacobian

$$D(\psi\varphi^{-1}): U \to GL(K)$$

where GL(K) is the general linear group over the reals of a Banach space isomorphic to $E^{r,s}$. Let $GL_{B_o}(E^{r,s})$ denote the B_o linear automorphisms of $E^{r,s}$. There is an inclusion

$$\theta: GL_{B_o}(E^{r,s}) \to GL(K).$$

The starred equality in 2.5.6 holds if the image of $D(\psi\varphi^{-1})$ lies in the image of θ. This is exactly what the condition for JP smoothness requires.

For the G^∞ case, let

2.5.7 $$M_{r,s}(B) = \left\{ \begin{bmatrix} A & B \\ C & D \end{bmatrix} : \begin{array}{l} A \text{ is an } r \times r \text{ matrix, entries in } B_o \\ B \text{ is an } r \times x \text{ matrix, entries in } B_1 \\ C \text{ is an } s \times r \text{ matrix, entries in } B_1 \\ D \text{ is an } s \times s \text{ matrix, entries in } B_o \end{array} \right\}$$

Obviously there is a map $\kappa: M_{r,s}(B) \to GL_{B_o}(E^{r,s})$. The G^∞ smoothness condition says that there is a map $D(\psi\varphi^{-1}): U \to M_{r,s}(B)$ which makes the following diagram commute:

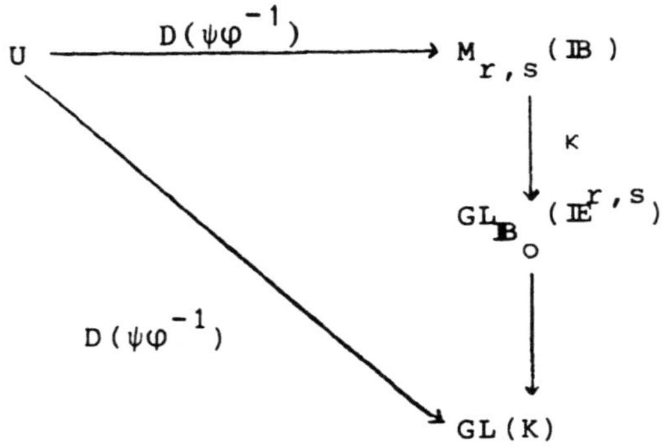

Hence the starred equality holds in the G^∞ case as well.

Finally the map

$$E^{r,s} \to T_{(x,y)} E^{r,s}$$

given by $(u,v) \to [c_{(u,v)}(t)]$ where

$$c_{(u,v)}(t) = (x_1+tu_1,\ldots,x_r+tu_r,y_1+tv_1,\ldots,y_s+tv_s)$$

provides a B_o linear isomorphism between T_sS and $E^{r,s}$.

The conjecture is that imitation of the methods for ordinary manifolds will provide the set $TS = \cup T_sS$ with the structure of a super vector bundle over S. In the JP case, see Jadzyck, Pilch [8].

2.5.8 Conjecture.

The set of JP smooth (G^∞) sections of TS have the structure of an ordinary Lie algebra.

Certainly, since a JP (G^α) section of TS is a C^∞ section of the ordinary tangent bundle of the ordinary smooth manifold S, and hence an element of the Lie algegra of derivations of functions. The question is whether the real linear subspace of JP (G^∞) sections is closed under the Lie bracket.

One would like to see Lie superalgebras arising naturally as sections of some analogue of tangent bundle. In the case of JP or G^∞ manifolds if $L = \infty$, Lie superalgebras seem to occur as sections of the tangent module, defined below.

2.5.9 Definition.

Let S be JP or G^∞ and let X be an S supermanifold. Define

$$T^{\sim}_sX(U) = \{\sigma : S(U) \to B : \sigma \text{ is B linear}$$
$$\sigma(fg) = \sigma(f)g(s) + (-1)^{|\sigma,f|} f(s)\,\sigma(g)$$
$$\text{for } f,g \text{ in } S(U)\}$$

for a coordinate neighbourhood U of s in X. If V is another coordinate neighbourhood of s, the restriction homomorphisms induce B linear maps

$$T^{\sim}_sX(U) \leftarrow T^{\sim}_sX(U\cap V) \to T^{\sim}_sX(V).$$

These maps can be shown to be isomorphisms (use the fact S(U) is isomorphic to $B \otimes C^\infty(\varepsilon(U)) \otimes \Lambda R^S$). Thus one

$$T^{\sim}_sX = T^{\sim}_sX(U)$$

where U is some coordinate neighbourhood of U and

$$T^{\sim}X = \bigcup T^{\sim}_{s}X.$$

2.5.10 Conjecture.

i) $T^{\sim}_{s}X$ is a free B module on r even generators and s odd generators.

ii) $T_{s}X \cong (T^{\sim}_{s}X)_{o}$ as B_{o} modules.

iii) $T^{\sim}X$ is a free B module bundle over X.

Proof.

i) Use the fact that $S(U) \cong B \otimes C^{\infty}(\varepsilon(U)) \otimes \Lambda R^{s}$.

ii) In the case that S = JP, see Jadzyck, Pilch [8].

iii) Generalize the usual argument that ordinary vector bundles correspond to locally free modules over the sheaf of functions. Now construct a sheaf M of free S modules as follows.

Define

$$M(U) = \{m\colon U \to T^{\sim}X\colon m(u) \text{ is in } T^{\sim}_{u}X$$
$$\text{the map } m_{g}\colon U \to B \text{ given by}$$
$$m_{g}(u) = \langle m(u), g\rangle$$
$$\text{is in } S(U) \text{ for all } g \text{ in } S(U)\}.$$

Section 3. Comparisons

The presence of different categories of supermanifolds reflects the fact that there are many basic theorems in differential topology which allow one to describe manifolds and constructions on manifolds in many different ways. (Think of the various ways of defining the ordinary tangent bundle.) In generalizing these methods to the graded commutative setting, each differential approach produces a genuinely different construction.

In the light of such differences I do not think it is likely that a single category of supermanifold will ever suit all purposes. The best that can be done is to accept the different categories and understand to what extent they can be compared.

3.1 Comparing GM and SSM.

This is the subject of Batchelor [2]. The basic idea is summarized here.

3.1.1 Step 1.

Create a functor

$$S: GM \to SSM.$$

This means, describe a way of starting with a GM graded manifold (X,A) and producing an SSM supermanifold S(X,A) such that if

$$f: (X,A) \to (Y,B)$$

is a map of graded manifolds, there is a map

$$Sf: S(X,A) \to S(Y,B).$$

This is done by defining

$$S(X,A) = \mathrm{Alg}(A(X),B)$$
$$= \{\text{algebra homomorphisms } s: A(X) \to B\}.$$

One must suppose that if $B = \Lambda R^L$ then L is greater than the odd dimension of (X,A). If f: (X,A) → (Y,B) is a GM map then f is an algebra homomorphism f: B(Y) → A(X) and f induces a map

$$Sf: S(X,A) = \mathrm{Alg}(A(X),B) \to \mathrm{Alg}(B(Y),B) = S(Y,B)$$

by composition with f.

The problem is to show that S(X,A) is an SSM supermanifold and that Sf is an SSM map.

3.1.2 Step 2.

construct a functor

$$G: SSM \to GM.$$

The method here is to define a nice manifold $\underline{S}$ given an SSM manifold S. This is done by setting

3.1.3 $\underline{S} = S/\sim$

where $s \sim s'$ if there exists a chart (U,φ) on S with s, s' in U and $\varepsilon\varphi(s) = \varepsilon\varphi(s')$.

3.1.4 Lemma.

The set S defined in 3.1.3 is a nice manifold.

Proof. (Sketch)

First check that ~ is an equivalence relation. Symmetry and the reflexive property are easy. Check transitivity. If s, s' are in a chart (U,φ), and s', s" are in a chart (V,ψ) then s' is in $U\cap V$. The claim is that s and s" are also in $U\cap V$. This is true because

$$\varphi^{-1}\varepsilon^{-1}(\varepsilon(\varphi(s'))) = \psi^{-1}\varepsilon^{-1}(\varepsilon(\psi(s')))$$

and s is in $\varphi^{-1}\varepsilon^{-1}(\varepsilon(\varphi(s')))$ and s is in $\psi^{-1}\varepsilon^{-1}(\varepsilon(\psi(s')))$.

Put a topology on $\underline{S}$ by saying that the open sets are the images of open sets in S. Clearly every point has a neighbourhood homeomorphic to an open set in R^r. To check that the transition maps are smooth, use the fact that if $f: \varepsilon^{-1}(V) \to B$ is an SSM map then the composition

$$V \to \varepsilon^{-1}(V) \xrightarrow{f} B \xrightarrow{\varepsilon} R$$

is a C^∞ function.

Now G can be defined.

3.1.5 Definition.

If S is an SSM supermanifold, define $\underline{S}$ as in 3.1.1. Let

$$\varepsilon_S: S \to \underline{S}$$

denote the projection. Define a sheaf A on $\underline{S}$ by setting

$$A(U) = SSM(\varepsilon_S^{-1}(U)).$$

Define $GS = (\underline{S},A)$.

If $f: S \to T$ is a map of SSM supermanifolds, then f induces an algebra homomorphism

$$Gf: SSM(T) \to SSM(S)$$

and $Gf: GS \to GT$ is a GM map.

The theorem that is proved is the following.

3.1.6 Theorem.

The functors G and S establish an equivalence between the category of graded manifolds of odd dimension less than L and the category of SSM supermanifolds for $B = \Lambda R^L$, $L<\infty$.

A more direct correspondence is sketched in section 3.3 in the case that the sheaf A is given as the sheaf of sections of an exterior bundle.

3.2 Comparison of geometric manifolds.

The problem of comparing geometric supermanifolds is much easier. Just as categories of supermanifolds could be defined by specifying a supereuclidean space and a criterion for smoothness, so different geometric categories can be compared by comparing the supereuclidean spaces and the sets of smooth maps.

3.2.1 Proposition.

Let S, S' be geometric categories of supermanifolds given by specifying graded commutative algebras B, B' with topologies τ, τ', and sets of smooth functions S, S'. Then S is a subcategory of S' if

i) $B = B'$

ii) $\tau < \tau'$

iii) If U is an open set in $E_B^{r,s} = E_{B'}^{r,s}$ with respect to τ, then $S(U) < S'(U)$.

The proof is not difficult, and the following corollary follows from comparisons of topologies and smoothness criteria given in section 2.

3.2.2 Corollary.

i) SSM is a subcategory of SM.

ii) SM is a full subcategory of G^∞. That is, if S, S' are SM supermanifolds and f: S → S' is a G^∞ map then f is an SM map.

iii) In the case that $B = \Lambda R^\infty$, G^∞ is isomorphic to JP.

3.2.3 Conjecture.

If S is an SM supermanifold, then S can be given the structure of an SSM supermanifold.

Idea of proof.

The idea is to start with an atlas (U_i, φ_i) of SM charts for S and use it to build an atlas $(U_i^1, \varphi_i^1{}')$ of SSM charts. Write

$$W_i = \varphi_i U_i$$

$$W_{i,j} = \varphi_i (U_i \cap U_j)$$

$$\tau_{ij} = \varphi_i \varphi_j^{-1} : W_{j,i} \to W_{i,j} .$$

By the definition given for SM smooth maps one can choose

$$\underline{\tau}_{ij} \in Q = (B \otimes C^\infty(\varepsilon W_{j,i}) \otimes \Lambda R^s)_o{}^r \times (B \otimes C^\infty(\varepsilon W_{j,i}) \otimes \Lambda R^s)_1{}^s$$

with $\underline{\tau}_{ij} = \tau_{ij}$ as functions on $W_{j,i}$. The map

$$\varepsilon_B : B \otimes C^\infty(\varepsilon W_{j,i}) \otimes \Lambda R^s \to C^\infty(\varepsilon W_{j,i}) \otimes \Lambda R^s$$

allows one to define

$$\underline{\sigma}_{ij} = \varepsilon_B \underline{\tau}_{ij} \in Q.$$

The $\underline{\sigma}_{ij}$ give rise to SSM maps $\sigma_{ij}: W_{i,j} \to W_{i,j}$. Check that the $\underline{\sigma}_{ij}$ and hence the σ_{ij} are independent of $\underline{\tau}_{ij}$. One can use the maps σ_{ij} to build an SSM supermanifold S',

$$S' = \cup W_i / \sim$$

where $w \sim y$ if w is in $W_{i,j}$ and y is in $W_{j,i}$ and $\sigma_{ij} y = w$. The conjecture is that S and S' are SM diffeomorphic.

3.3 A direct method of comparing GM and G^∞.

The method described in 3.1 may establish the correspondence between GM and SSM, but it fails to provide a satisfactory intuitive geometric idea of what the process is. Such an intuitive grasp is necessary to work effectively with supermanifolds.

In the case that a graded manifold (X,A) is given by describing A as the sections of an exterior bundle over X, the corresponding geometric supermanifold can be constructed directly.

Throughout this section let $\xi: E \to X$ be an ordinary vector bundle over a nice manifold X. Let A be the sheaf of smooth sections of ΛE.

3.3.1 Step 1.

Given X of dimension r create a G^∞ supermanifold S of dimension (r,0).

Choose an atlas $(U_\alpha, \varphi_\alpha)$ for X. Set

$$W_\alpha = \varphi_\alpha U_\alpha$$
$$W_{\alpha,\beta} = \varphi_\alpha(U_\alpha \cap U_\beta)$$
$$V_\alpha = \varepsilon^{-1} W_\alpha < E^{r,o}$$
$$V_{\alpha,\beta} = \varepsilon^{-1} W_{\alpha,\beta} < V_\alpha .$$

Define

$\tau_{\alpha\beta}: V_{\beta,\alpha} \to V_{\alpha,\beta}$ to be the map extended from

$\varphi_\alpha \varphi_\beta^{-1}: W_{\beta,\alpha} \to W_{\alpha,\beta}$.

(Use the map including $C^\infty(\varepsilon(U)) \to G^\infty(U)$ for an open set U in $E^{r,s}$.) Then define

$$S = \cup V_\alpha / \sim$$

where $v \sim u$ if v is in $V_{\alpha,\beta}$ and u is in $V_{\beta,\alpha}$ and $\tau_{\alpha\beta} u = v$. Check that S is a G^∞ supermanifold.

3.3.2 Step 2.

Use E to build an (r,s) dimensional G^∞ supermanifold T with a G^∞ map $p: T \to S$.

Notice that there is a map of nice (ordinary) manifolds

$$\varepsilon: S \to X$$

using the construction in 3.1.3. Let E' be the dual bundle of E over X, and define a new bundle $\rho: F \to X$ by setting $F = E' \otimes B_1$. Finally define T to be the pull-back of F via ε

$$\begin{array}{ccc} T & \longrightarrow & F \\ \downarrow & & \downarrow \rho \\ S & \underset{\varepsilon}{\longrightarrow} & X \end{array}$$

That is, T is the subset of elements (s,f) in $S \times F$ satisfying $\varepsilon s = \rho f$.

3.3.3 Conjecture.

The supermanifold T correspond to (X,A) under the correspondences described in sections 3.1 and 3.2.

There is a map

$$q: T \to Alg(A,B)$$

as follows. Let (s,f) be in T with $\rho f = \varepsilon s = x$. Let σ be in $A(U) = \Gamma(U,\Lambda E)$. Let $e_1,\ldots,e_s$ be sections of E over U such that the $e_j(x)$ form a basis for E_x. Let $\xi_1,\ldots,\xi_s$ be the dual basis in E'_x. Let $\nu = (j_1,\ldots j_s)$ be an s tuple of elements j_k in $\{0,1\}$. Write

$$e(\nu) = e_1^{j_1}\wedge\ldots\wedge e_s^{j_s}$$

$$\sigma = \Sigma\, r_\nu e(\nu).$$

$$f = \Sigma\, \xi_j b_j$$

$$b(\nu) = b_1^{j_1}\wedge\ldots\wedge b_s^{j_s}$$

$$\underline{r_\nu} = \text{the image of } r_\nu \text{ in } F(\varepsilon^{-1}U)$$

and set

$$q(s,f) = \Sigma\, \underline{r_\nu}(s)b(\nu).$$

To see that q is a bijection, appeal to the following lemma.

3.3.4 Lemma.

$Alg(C^\infty(V) \otimes \Lambda R^s, B) = E^{r,s}$ if $B = \Lambda R^L$ with $s < L < \infty$.

Proof.

See the definition of SSM smooth maps for the inclusion of $C^\infty(V) \otimes \Lambda R^s$ into the set of functions on $\varepsilon^{-1}(V)$. This inclusion is an algebra homormorphism. (See Batchelor [2]). Hence elements of $\varepsilon^{-1}(V)$ act as algebra homomorphisms from $C^\infty(V) \otimes \Lambda R^s$ to B.

Conversely, if k is an algebra homomorphism it can be shown that

$$(k(u_1),\ldots,k(u_r),k(v_1),\ldots,k(v_s))$$

where the u_j are coordinate projections for V and the v_j generate ΛR^s, describes a point in $\varepsilon^{-1}(V)$.

Section 4. Lie supergroups.

4.1 Lie supergroups in the geometric categories.

4.1.1 Definition.

Let S be SSM, SM, G^∞, G^ω, or JP. An S Lie supergroup G is a topological group G such that G is an S supermanifold and the operations of multiplication and taking inverses are S smooth.

4.1.2 Remarks.

i) For S = G^∞ or JP, an S Lie supergroup is an ordinary Lie group. In particular, using the definition of tangent space given in 2.4.4, the tangent space at the identity can be given the structure of an ordinary Lie algebra. As such, the exponential map is legitimate.

ii) Let G be a JP or G^∞ with $L = \infty$ Lie supergroup. The action of G on itself gives an action

$$G \times T^{\sim}G \to T^{\sim}G$$

given by $g\gamma(f) = \gamma(f \circ g)$, for $g \in G$, f a JP (G^∞) function, and γ in $T^{\sim}G$. Let e be the identity in G. Using this action, elements of $T_e{}^{\sim}G$ yields a section of $T^{\sim}G$, that is, a derivation of JP(G) (or $G^\infty(G)$). These derivations appear to form a Lie superalgebra.

4.1.3 Examples.

Let $s = 2k$, and let F be the r+s square matrix

$$F = \left[\begin{array}{c|cc} I_r & \multicolumn{2}{c}{0} \\ \hline 0 & 0 & I_k \\ & -I_k & 0 \end{array}\right]$$

Let $M_{r,s}(B)$ be the set of r+s square matrices with entries in B defined in 2.5.6. Define the supertranspose of a block matrix by

$$^{T}\begin{bmatrix} A & B \\ C & D \end{bmatrix} = \begin{bmatrix} {}^tA & {}^tC \\ -{}^tB & {}^tD \end{bmatrix}$$

where ${}^tY$ is the ordinary transpose of a matrix Y. Let G be the subgroup of matrices given by

$$G = \{X \in M_{r,s}(B) : {}^{T}XFX = F\}.$$

Since G is a subset of $M_{r,s}(B) = E^{r+s,2r}$ which is a topological space (with the fine topology), G is also a topological space. Check that G is a topological group.

Consider the subset L of matrices A in $M_{r,s}(B)$ satisfying

$${}^{T}AF + FA = 0.$$

These form a B_o submodule and an ordinary Lie algebra. The claim is that $\mathcal{L}$ is a vector superspace.

Consider F as a form on a real r+s dimensional vector space V., with F symmetric on V_o and F skew symmetric on V_1. Then End(V) is a Z_2 graded associative algebra: it can be considered as a Lie superalgebra by setting

$$[A,B] = AB + (-1)^{|A,B|}BA.$$

The linear subspace

$$M = \{A \text{ in } End(V) : {}^{T}AF + FA = 0\}$$

is a Lie superalgebra over the reals, called the orthosympletic Lie superalgebra. This also means that M × B is a Lie superalgebra and

$$(M \otimes B)_o = M_o \otimes B_o + M_1 \otimes B_1 = L .$$

Thus L is a vector superspace as desired. See Scheunert [14].

Now check that the exponential map sends a neighbourhood of 0 in L homeomorphically onto a neighbourhood U of the identity of G. Thus $(U, \exp^{-1})$ is a chart at the identity and $(gU, \exp^{-1}g^{-1})$ is a chart at g. The transition maps are G^{ω} and hence G^{∞}. This shows that G is a G^{∞} Lie supergroup, and gives some insight into the relationship between graded Lie algebras and Lie supergroups.

ii) Let $B = \Lambda R^1$. Addition in $B = E^{1,1}$ gives B the structure of a G^{∞} Lie supergroup. Notice that the torus constructed in 2.4.5 is the quotient of B by a discrete subgroup, and hence a G^{∞} Lie supergroup. It is an example of a compact Lie supergroup.

iii) Notice that $E^{r,s}$ is not a Lie supergroup in the category SSM.

4.2 Graded Lie Groups

4.2.1 The idea.

Recall from 1.1.6 that if $X = R$ the dual coalgebra $C^\infty(X)^o$ is given as the linear space

$$C^\infty(X)^o = \left\langle \frac{d}{dt}\bigg|_p : i = 0,1,2,\ldots,\ p \in X \right\rangle.$$

Another way of writing this is

$$C^\infty(X)^o = RX \otimes U(L)$$

where $R(X)$ is the vector space spanned by points in X and $U(L)$ is the universal enveloping algebra of the one dimensional Lie algebra L corresponding to the Lie group $X = R$.

In general, if G is any finite dimensional Lie group, $C^\infty(G)^o$ is a Hopf algebra

$$C^\infty(G)^o = RG \otimes U(L)$$

where $U(L)$ is the universal enveloping algebra of the Lie algebra L of the group G. Each of RG and $U(L)$ is a Hopf algebra by defining multiplication by

4.2.2 $$(g \otimes x)(h \otimes y) = (gh \times [ad(h)x,y]) \quad g,h \in G,\ x,y \in L$$

where $ad: G \times L \to L$ is the adjoint action.

These properties can be generalized to the graded commutative setting to give a definition of groups for GM.

4.2.3 Definition-proposition. Lie-Hopf algebras. (See Kostant [9]).

Let **L** be a Lie superalgebra and let G be a Lie group with Lie algebra L_o. Suppose G acts on L such that the induced action of L_o on L is the adjoint action. Then

$$H = RG \otimes U(L)$$

can be given the structure of a Hopf algebra, with multiplication given by 4.2.2. Such a Hopf algebra will be called a Lie-Hopf algebra.

Since (by 4.2.1) $C^\infty(G)^o$ is a Lie-Hopf algebra, it is reasonable to make the following definition.

Categories / points of comparison	ALGEBRAIC APPROACH idea: You only need to know the functions. Berezin, Kostant GM	idea: To get anti-commuting functions Common Properties
DEFINING PROPERTY: A supermanifold is to be an analogue of manifold the function ring of which contains anti-commuting elements	Definition (1.1.1) A graded manifold is a pair (X,A) X - nice manifold A - sheaf of $\mathbb{Z}_2$ graded commutative algebras with (i) An onto map of sheaves of algebras $\varepsilon: A \to C^\infty$. (ii) An open cover U_i of X and isomorphisms of sheaves $\tau_i\ A\|_{U_i} \to C^\infty(\ ) \otimes \Lambda \mathbb{R}^s\|_{U_i}$ Examples (1.1.2) (X,C^∞), $(*,\Lambda\mathbb{R}^s)$ (X,Ω) Ω is the sheaf of forms on X $(X,\Gamma(\ ,\Lambda E))$, $\Gamma(\ ,\Lambda E)$ is the sheaf of sections of the exterior bundle of a vector bundle E over X. Point set. Use the dual coalgebra $A(X)^o$ for an analogue of point set. (1.1.5, 1.1.6).	To define a geometric supermanifold you need (2.1.1) $\mathbb{B}$ - graded commutative algebra topology on $\mathbb{B}$ (in fact, all examples have an augmentation map $\varepsilon: \mathbb{B} \to \mathbb{R}$. $\mathbb{E}_{\mathbb{B}}^{r,s} = (\mathbb{B}_o)^r \times (\mathbb{B}_1)^s$ supereuclidean space, with the topology inherited from the topology on $\mathbb{B}$. Smoothness criterion: A way of recognizing the set of smooth functions S(U,V) in the set of all functions F(U,V) for U,V appropriate open sets. Write S(U) for S(U,B).
BUNDLES:	Definition (1.2.1) A vector bundle over (X,A) is a locally free sheaf of A modules. That is, define a bundle by describing the sections. Proposition (1.2.4) A vector bundle M over (X,A) gives rise to an ordinary vector bundle E_M over X. Tangent Bundle (1.2.2i) Make the set Der(U,A) of derivations of A(U) into a sheaf. $E_{Der(\ ,A)}$ is the vector bundle T(X,A) over X consisting of differentiations from A(X) to $\mathbb{R}$. Forms (1.2.2ii) Define $\Omega(U,A)$ to be the $Z \times Z_2$ graded commutative algebra freely generated by elements a, db with a, b in A(U) subject to the condition that d is a derivation of degree (I,0).	A vector superspace V(2.5.2) is a $\mathbb{B}_0$ module isomorphic to $E^{r,s}$. Convention: $V = V_0$. A vector bundle (2.5.5) $\xi: F \to S$ is an S smooth map between S-supermanifolds such that $\xi'(s)$ is a vector superspace for all s in S, and s has a neighbourhood U such that F is locally isomorphic to the product $U \times \mathbb{E}^{r,s}$. If $\pi: H \to S$ is a vector bundle such that $\pi^{-1}(s)$ is a free $\mathbb{B}$-module, then $\pi: H \to S$ is called a free $\mathbb{B}$ module bundle.
GROUPS: Examples Correspondence with Lie superalgebras.	Definition (4.2.4) A graded Lie group is a graded manifold (G,A) such that the coalgebra $A(G)^o$ is a Hopf algebra isomorphic to $\mathbb{R}G \otimes U(L)$. Example (4.2.6) The orthosymplectic group. The tangent space at the identity is a Lie superalgebra (4.2.3).	Definition (4.1.1) A Lie supergroup is a group G which has the structure of a supermanifold where the operations of multiplication and taking inverses are smooth.
COMPARISONS:	GM	Theorem (1.1.3, 1.2.8) conjecture (3.3)
CLASSIFICATION THEOREMS:	Theorem A is isomorphic to a sheaf of sections of an exterior bundle E over X (1.7.3). Theorem (1.2.8) The bundle E is dual to the odd tangent bundle $T(X,A)_1$.	

GEOMETRIC APPROACH

, start with an anti-commuting algebra.

Batchelor SSM	De Witt SM	Rogers $G^\infty(G^\omega)$	Jadzyck Pilch JP
$\mathbb{B} = \Lambda\mathbb{R}^L$, $L < \infty$, $\mathbb{B} = \Lambda\mathbb{R}^L$, $L < \infty$		$\mathbb{B} = \mathbb{R}^L$, $L < \infty$ with Banach algebra structure	$\mathbb{B}$ is an augmented Z_2 graded commutative Banach algebra satisfying

Batchelor / De Witt

$\mathbb{B} = \Lambda\mathbb{R}^L$, $L < \infty$, $\mathbb{B} = \Lambda\mathbb{R}^L$, $L < \infty$

Coarse topology: U is open in $\mathbb{B}$ $\Leftrightarrow$ $U = \varepsilon^{-1}(V)$ for V open in $\mathbb{R}$.

Algebraic definition (2.3.6), (2.3.7) S(U) is the subalgebra of functions in F(U) containing elements of the following form.

I. Coordinate projections and polynomials (over $\mathbb{R}$) in these projections.

II. Extensions of $\mathbb{R}$-valued smooth functions on (U).

III. (SM only) Constant $\mathbb{B}$-valued functions.

$SSM(U) = C^\infty(\varepsilon U) \otimes \Lambda\mathbb{R}^s$

$\mathbb{B} \otimes C^\infty(\varepsilon U) \otimes \Lambda\mathbb{R}^s$ maps onto SM(U).

Rogers

$\mathbb{B} = \mathbb{R}^L$, $L < \infty$ with Banach algebra structure

Definition (2.3.1) A continuous function $f: U \to \mathbb{B}$ is G^1 if there exist continuous "partial derivatives"

H_i $\quad i = 1,\ldots,r$
K_j $: U \to \mathbb{B}$ $\quad j = 1,\ldots,s$

and a continuous map

$$\nu: U \times \mathbb{E}^{r,s} \to \mathbb{B}$$

such that for (u,v), (u+h,v+k) in U

$f(u+h,v+k) = f(u,v) + \Sigma h_i H_i(u,v) + \Sigma k_j K_j(u,v) + \nu((u,v),(h,k))$

and $\|\nu\| \to 0$ as $\|(h,k)\| \to 0$. Define G (u) inductively. G (U) is the set of analytic functions on U (2.3.2).

There is an onto map

$\mathbb{B} \otimes C^\infty(\varepsilon(U) \otimes \Lambda\mathbb{R}^s \to G^\infty(U)$

which is an isomorphism if $L = \infty$.

Jadzyck Pilch

$\mathbb{B}$ is an augmented Z_2 graded commutative Banach algebra satisfying

(i) The map $\mathbb{B}_j \quad L_{\mathbb{B}_0}(\mathbb{B}_i, \mathbb{B}_{i+j})$ is a Norm preserving isomorphism .

(ii) For b in $\mathbb{B}_0$ $\|b\| = \|b - \varepsilon(b)\| + (\varepsilon(b))$

Example: $\Lambda\mathbb{R}^\infty$,

Definition (2.3.4) A continuous function $f: U \quad \mathbb{B}$ is JP' if there exist continuous functions

$G: U \to GL_{\mathbb{B}_0}(\mathbb{E}^{r,s}, \mathbb{B})$

$\eta: U \to \mathbb{B}^{r,s} \to \mathbb{B}$

such that if u, u+h are in U, then

$f(u+h) = f(u) + G(u)h + \eta(u,h)$

and $\|\eta\| \to 0$ as $\|h\| \to 0$. Define $JP(U)^P$ by induction and say f is JP if f is JP^P for all P:

$JP(U) \cong \mathbb{B} \otimes C^\infty(\varepsilon(U)) \otimes \Lambda\mathbb{R}^s$.

"ALL I KNOW

ABOUT

SUPERMANIFOLDS"

Marj. Batchelor

July, 1983

The tangent bundle (2.5.6). For both these categories, supermanifolds are smooth manifolds with additional structure. Define the tangent space T_sS at a point s as usual, to be the set of equivalence classes of curves at s. Form TS as usual.

The tangent module (2.5.9) (JP or G^∞, $L = \infty$) The set of differentiations $JP(U) \to \mathbb{B}$ ($G^\infty(U) \to \mathbb{B}$) at a point s is a free $\mathbb{B}$ module $T^\sim_sS$. Set $T^\sim S = \bigcup_s T^\sim_sS$.

There is an inclusion $TS \subset T^\sim S$ (2.5.10).

Counter example: $\mathbb{E}^{r,s}$	Example: $\mathbb{E}^{r,s}$	Example (4.1.3) The group of matrices in $M_{r,s}(\mathbb{B})$ preserving a bilinear form F. (The orthosymplectic group). Example (2.4.5) G^∞ only. The torus group. The tangent space at the identity is an ordinary Lie algebra. The tangent module at the identity is a Lie superalgebra.

$\longrightarrow$ SSM $\xleftarrow{\text{conj.(3.2.3)}}$ SM $\longrightarrow$ G^∞ $\qquad$ $G^\infty_{L=\infty}$ $\xleftrightarrow[\text{3.2.2}]{\text{conj.}}$ JP

Use the correspondence of SSM with GM to classify SSM	See conjecture (3.2.3)	STOP PRESS: For G^∞ supermanifolds this problem has been analysed in a paper by Boyer and Gitler (Prep.)

4.2.4 Definition.

A graded Lie group is a graded manifold (G,A) where the coalgebra $A(G)^o$ has the structure of a Lie-Hopf algebra.

4.2.5 Corollary.

Let (G,A) be a graded Lie group and let L be the Lie superalgebra such that $A(G)^o = RG \otimes U(L)$. If V is the dual space of L_1, then

$$A \cong C^\infty \otimes \Lambda V.$$

Proof.

Use theorem 1.2.9 and then observe that the odd tangent space $T(G,A)_1$ must be trivial: Take a basis $(\gamma_1,\ldots,\gamma_s)$ for $T_e(G,A)_1 = L_1$ and observe that the sections φ_i of $T(G,A)$ given by

$$\varphi_i(g) = g\gamma_i$$

are linearly independent at each point.

4.2.6 Example.

Let M be the orthosymplectic Lie superalgebra defined in 4.1.3. This is a subspace of the set of $(r+s)\times(r+s)$ matrices. The group $G = O(r) \times Sp(s)$ preserves the form F, and acts on elements of M by conjugation. Thus $RG \times U(M)$ is a Lie-Hopf algebra. Define

$$A = C^\infty \otimes \Lambda(M_1').$$

REFERENCES

[1] Batchelor, M. *The structure of supermanifolds*. Trans. Amer. Math. Soc. 253 (1979) 329-338.

[2] Batchelor, M. *Two approaches to supermanifolds*. Trans. Amer. Math. Soc. 258 (1980) 257-270.

[3] Batchelor, M. *An algebraic approach to integration*. Preprint.

[4] Batchelor, M. *A functor from graded manifolds to vector bundles*. In preparation.

[5] Berezin, F. *Differential forms on supermanifolds*. Sov. J. Nucl. Phys.

[6] Berezin, F., Leites, D. *Supervarieties*. Soviet Math. Dokl. 16 (1975) 1218-1222.

[7] deWitt, B. *Differential supergeometry*. Preprint.

[8] Jadzyck, A., Pilch, K. *Superspaces and supersymmetries*. Comm. Math. Phys. 78 # 3, (1981) 373-390.

[9] Kostant, B. *Graded manifolds, graded Lie theory and prequantization*. Differential Geometric Methods in Mathematical Physics, Lecture Notes in Mathematics, 570, Springer, (1977), 177-306.

[10] Leites, D. *Russian Math. Surveys* 35 : 1 (1980) 1-64.

[11] Rogers, A. *A global theory of supermanifolds*. J. Math. Phys. 21 # 6 (1980) 1352-1365.

[12] Rogers, A. *Some examples of compact supermanifolds with non-abelian fundamental group*. J. Math. Phys. 22 # 3 (1981) 443-444.

[13] Rogers, A. *Super Lie groups: global topology and local structure*. J. Math. Phys. 22 # 5 (1981) 939-945.

[14] Scheunert, M. *Lie superalgebras*. Lecture Notes in Mathematics 716, Springer, 1979.

[15] Sweedler, M. *Hopf algebras*. Benjamin, New York. (1980).

[16] Wells, R.O. *Differential analysis on complex manifolds*. Graduate Texts in Mathematics 65, Springer, New York, (1980).

ASPECTS OF THE GEOMETRICAL APPROACH TO SUPERMANIFOLDS

Alice Rogers

Department of Mathematics, King's College,
Strand, London WC2

Research supported by the SERC

Abstract Various topics in the theory and application of the geometrical approach to supermanifolds are discussed. The construction of the superspace used in supergravity over an arbitrary spacetime manifold is described. Super Lie groups and their relation to graded Lie algebras (and more general structures referred to as 'graded Lie modules') are discussed, with examples. Certain supermanifolds, allowed in the geometric approach (using the fine topology), but having no analogue in the algebraic approach, are discussed. Finally lattice supersymmetry, and its relation to the differential geometry of supermanifolds, is discussed.

In this article I shall consider various topics in the theory and application of supermanifolds from the geometric point of view; the relation between the algebraic definition of "graded manifold" [1,2] and the geometric definition of "supermanifold" [3,4,5,6] has been described in Batchelor's lectures [7]; her lectures emphasise the use of algebraic-geometric techniques and are an excellent introduction for physicists to this less familiar approach. I shall therefore concentrate on the geometric approach, discussing in particular some specific structures used in theoretical physics, and some supermanifolds for which there are no analogous graded manifolds. In section 2 the superspace used in supergravity theories is described, section 3 contains a discussion of super Lie groups and their relation to graded Lie algebras, section 4 contains examples of supermanifolds with exotic topology in the odd directions, while section 5 looks at lattice supersymmetry and its relation to certain mathematical problems for supermanifolds.

H.-J. Seifert et al. (eds.), Mathematical Aspects of Superspace, 135–147.

Throughout this article, the supermanifolds will be "G^∞" supermanifolds [5,7]; this means that local coordinates on an (r,s) dimensional supermanifold have the form $(x^1,x^2,\ldots,x^r;\ y^1,y^2,\ldots,y^s)$ with the x^i taking values in the even part of a Grassmann algebra and the y^j in the odd part. The Grassmann algebra may be B_L, the Grassmann algebra over a finite number, L, generators $\beta_1,\ldots,\beta_L$. (For simplicity, attention will be confined to Grassmann algebras over the reals, though other fields can easily be used.) The norm (and hence the topology) on B_L is defined by

$$\left\|a_o + \sum_{i=1}^{L} a^i \beta_i + \sum_{\substack{i,j=1\\ i<j}}^{L} a^{ij} \beta_i \beta_j + \ldots + a^{1\ldots L} \beta_1 \cdots \beta_L\right\|$$

$$= |a_o| + \sum_{i=1}^{L} |a^i| + \sum_{\substack{i,j=1\\ i<j}}^{L} |a^{ij}| + \ldots + |a^{1\ldots L}|. \tag{1.1}$$

This definition can be extended to the case $L \to \infty$, to give B_∞, if one additionally imposes the condition

$$\|b\| < \infty \qquad \forall\, b \in B_\infty . \tag{1.2}$$

Further details may be found in [5], where it is proved that B_L (for both finite and infinite L) forms a Banach algebra; so that as well as the useful properties of a Banach space, one has

$$\|1\| = 1$$
$$\text{and} \quad \|ab\| \leq \|a\|\,\|b\| . \tag{1.3}$$

This means that a definition of 'differentiation' of functions of Grassmann variables can be given in terms of changes in the function corresponding to changes in its argument, very much as in conventional differentiation [5,7]; the Leibnitz rule (in graded form), the rule for differentiating composite functions, and so on, are all valid. Functions infinitely differentiable in this sense are called "G^∞". For finite L, an alternative classification of G^∞ functions is possible: following Batchelor [7] let $E^{r,s}$ denote super Euclidean space, the cartesian product of r copies of the even part and s copies of the odd part of B_L; also let ε denote the "augmentation map" $\varepsilon\colon B_L \to \mathbb{R}$ which projects B_L onto $\mathbb{R}$ by sending the odd generators to zero; (one also has $\varepsilon_{r,s}\colon E^{r,s} \to \mathbb{R}^r$.) Then if U is an open set in $\mathbb{R}^r$, and $f\colon U \to B_L$ is C^∞ we can construct a function

$$z(f)\colon \varepsilon^{-1}(U) \to B_L$$

with $z(f)\ (x^1,\ldots,x^r;\ \ y^1,\ldots,y^s)$

$$:= f(\varepsilon(x^1,\ldots,x^r)) + \sum_{j=1}^{r} (x^j - \varepsilon(x^j))\ \partial_j\ f(\varepsilon(x^1,\ldots,x^r) + \ldots \tag{1.4}$$

(Because each $x^j - \varepsilon(x^j)$ is nilpotent, the Taylor series will terminate.) Using this z-map one can (as described also in [4,7]) define the set of G^∞ functions of ε^{-1} (U) into B_L to be the set of all functions of the form

$$g:\ \varepsilon^{-1}\ (U) \to B_L$$

$$g(x^1,\ldots,x^r;\ \ y^1,\ldots,y^s)$$

$$= z(f_o)\ (x^1,\ldots,x^r) + \sum_{j=1}^{s} z(f_j)\ (x^1,\ldots,x^r)\ y^j + \sum_{\substack{j_1,j_2=1\\ j_1<j_2}}^{s} z(f_{j_1j_2})\ (x^1,\ldots,x^r)\ y^{j_1}\ y^{j_2} + \ldots + z(f_{1\ 2\ldots s})\ (x^1,\ldots,x^r)\ y^1y^2\ldots y^s \tag{1.5}$$

for some f_o, f_j, $f_{j_1j_2}$, ..., $f_{1\ 2..s}$ in $C^\infty(U,B_L)$.

This definition coincides with the definition obtained by looking directly at variations in g corresponding to variations in $(x^1,\ldots,x^r;\ y^1,\ldots,y^s)$. (When L is finite there are some ambiguities in both approaches, basically because the product of all the generators $\beta_1\ \beta_2\ \ldots\ \beta_L$ is annihilated by multiplication with any single generator. This problem was one motivation for the introduction of B_∞; there may be other ways of allowing L to become infinite.)

The topology on B_L and $E^{r,s}$ will be the topology induced by the norm, which is the finer of the two topologies used in supermanifold theory.

2 BUILDING SUPERSPACE OVER AN ARBITRARY SPACETIME

The considerable complexities of calculations in supersymmetric theories are often most effectively handled by an extension of conventional spacetime known as "superspace" [8,9];

at the level of local coordinates one simply replaces the usual real x^m ($m = 0,\ldots,3$) by (x^m,θ^μ) ($\mu = 1,\ldots,n$) where the x^m now take values in the even part of B_L and the θ^μ in the odd part; if the spacetime is not simply $\mathbb{R}^4$ but is now general 4 dimensional manifold which cannot be covered by a single coordinate neighbourhood, the question of how one extends the manifold to be a (4,n) dimensional supermanifold is less trivial; the necessary construction is first to extend to a (4,0) dimensional supermanifold by extending the transition functions for change of coordinate on the manifold to the full even part of the Grassmann algebra B_L; the odd part is then included by a simple direct product. In detail, if M is a 4-dimensional real manifold with atlas $\{(U_\alpha,\phi_\alpha)\ \alpha \in \Lambda\}$(so that the U_α form an open cover by M, each ϕ_α maps U_α homeomorphically onto $\mathbb{R}^4$, and $\phi_\alpha \circ \phi_\beta^{-1}$ is a diffeomorphism of $\phi_\beta\ (U_\alpha \cap U_\beta)$ and $\phi_\alpha\ (U_\alpha \cap U_\beta)$ we can construct a (4,0) dimensional supermanifold N in the following way:

$$\text{let}\quad V_\alpha := \varepsilon^{-1}_{4,0}\ (\phi_\alpha(U_\alpha)) \subset E^{4,0} \tag{2.1}$$

$$\text{and}\quad \tau_{\alpha\beta}\colon \varepsilon^{-1}_{4,0}\ (\phi_\beta(U_\alpha \cap U_\beta)) \to \varepsilon^{-1}_{4,0}\ (\phi_\alpha(U_\alpha \cap U_\beta))$$

be defined (using the z-map of equation (1.4)) by

$$p_i \circ \tau_{\alpha\beta}(x^1,\ldots,x^r) = z(p_i \circ \phi_\alpha \circ \phi_\beta^{-1})\ (x^1,\ldots,x^r) \qquad i = 1,\ldots,r \tag{2.2}$$

(where p_i denotes projection onto the ith factor).

$$\text{Also let}\quad V = \underset{\alpha\in\Lambda}{\underline{\cup}}\ V_\alpha \tag{2.3}$$

(where $\underline{\cup}$ denotes disjoint union).

Next define on V the equivalence relation R by given p,q in V, pRq if and only if $\tau_{\alpha\beta}(p) = q$ where α,β are the unique elements of Λ such that $p \in V_\beta$ and $q \in V_\alpha$. N is defined to be the quotient space V/R. N has a natural supermanifold structure, by what might be called a 'z-extension' of the atlas of M: denoting equivalence classes of R by square brackets, define

$$\begin{aligned} &\psi_\alpha\colon [V_\alpha] \to V_\alpha \quad \text{by} \\ &\psi_\alpha([p]) := p \end{aligned} \tag{2.4}$$

(p being the unique element of [p] in V_α).

This construction (which is a special case of the construction described by Batchelor in [4,7] and by the author in [5], and is also described by Bonora, Pasti and Tonin [10],) shows directly how the coordinate patching of a non-trivial real manifold can be extended to the whole of the even part of B_L; the only problem would be with exotic transition functions when using B_∞. The (4,n) dimensional superspace is obtained simply as $N \times E^{0,n}$.

Another feature of the underlying curved spacetime which must be extended to superspace is the local Lorentz structure; on the spacetime manifold this is represented by a metric or equivalently by a reduction of the $GL(4,\mathbb{R})$ bundle of frames to an $SO(3,1)$ bundle. In the case of supergravity, it is this bundle reduction which is carried over to superspace - merely to put a metric on superspace does not give a tight enough structure [9]. As an example, for N = 1 supergravity one uses a (4,4) dimensional superspace; this has a frame bundle with structure group GL(4,4); in other words, one defines a frame to be an ordered basis of the tangent space, with the additional condition that the first four basis elements are even and the second four odd. Such bases are then transformed into one another by invertible (4,4) × (4,4) "supermatrices", that is to say, matrices of block form

$$M = \begin{pmatrix} A & X \\ Y & B \end{pmatrix} \tag{2.5}$$

where A,B,X and Y are 4 × 4 matrices with A and B having even Grassmann entries and X and Y having odd ones. In the superspace of N = 1 supergravity, one reduces from GL(4,4) to SL(2,C); the preferred frames are labelled $E_A = \{E_a, E_\alpha\}$ $a = 0,\ldots,3$ $\alpha = 1,\ldots,4$ with SL(2,C) action defined by the Majorana representation on the E_α and the vector representation on the E_a.

Because of the trivial way in which the odd part of the superspace is added on, the various further operations which are carried out remain possible. In particular, integrals over superspace can be defined, because the type of coordinate transition which makes this difficult [11] is not required. Also it is possible to retrieve the spacetime from the superspace, and to specify the partial gauge choice in superspace used to give the required physical fields, in a global manner [12]. Whether these things would be possible if one built the odd part over the even part in a less trivial way is an open question; and the physical significance of any such non-trivial 'super topology' has not been analysed. But it is at least clear that a superspace with the properties required for the conventional superspace techniques of supergravity can be "glued on" to any spacetime.

3 SUPER LIE GROUPS

Much current mathematical research in supermanifolds was inspired by a need to investigate in a rigorous manner structures first discovered in the context of supersymmetric quantum field theories. Super Lie groups are an example of this; initially these were looked at infinitesimally and heuristically - starting with a graded Lie algebra one simply combined odd generators with odd infinitesimal parameters (and even with even) and then obtained finite transformations by exponentiation (using the Baker-Campbell-Hausdorff formula). A full theory of super Lie groups, with many analogies to classical Lie group theory, has been developed and will now be summarised; some examples will be given. The odd and even nature of the parameters of the heuristic supergroup motivate the following definition [13]:

Definition 3.1

An (m,n) dimensional super Lie group is a group G which is also an (m,n) dimensional supermanifold such that the map

$$\begin{aligned} G \times G &\to G \\ (g_1, g_2) &\rightsquigarrow g_1 \circ g_2^{-1} \end{aligned} \tag{3.1}$$

is G^∞.

With classical Lie groups, the classification problem is largely handled by considering Lie algebras; a similar approach is taken here, the analogue of the Lie algebra being a "graded left B_L-module":

Definition 3.2

A graded Lie left B_L-module (GLM) is a graded Lie algebra W which is also a left B_L module such that

$$[bX_1, X_2] = b[X_1, X_2] \text{ for all } b \text{ in } B_L,\ X_1, X_2 \text{ in } W. \tag{3.2}$$

If the GLM is a free B_L-module with m even and n odd generators then it is said to have dimension (m,n); by considering expressions for one basis in terms of another one can show that the dimension is well-defined. A fairly straightforward generalisation of the classical proof of Lie's first theorem shows that the set $L(G)$ of left-invariant vector fields of an (m,n) dimensional super Lie group G form a GLM with bracket operation

$$\begin{aligned} [\ ,\ ] : L(G) \times L(G) &\to L(G) \\ (X,Y) \quad &[X,Y] := XY - (-1)^{|X||Y|}\, YX. \end{aligned} \tag{3.3}$$

(An important ingredient in the proof is Kostant's proof [2] that the vector fields on a graded manifold form a graded Lie algebra under graded commutation.)

Because B_L is a Banach space, and G^∞ maps are always C^∞, a supermanifold is a fortiori a manifold, and a super Lie group is always a Lie group. Moreover if G is a super Lie group with GLM denoted W, then it also has the structure of a Lie group with Lie algebra equal to the even part of W (regarded now simply as a Lie algebra). Thus in proving results about supergroups, one can harness results about classical Lie groups. In particular one has the equivalent of the converse of Lie's second theorem [13], at least when L is finite: given any GLM W of dimension (m,n), then any Lie group with Lie algebra equal to the even part of W (regarded simply as a Lie algebra) has also the structure of an (m,n) dimensional super Lie group with GLM equal to W. Thus the problem of classifying super Lie groups is reduced to the problem of classifying GLMs. It is often assumed that the GLM of a super Lie group is the tensor product of B_L with some graded Lie algebra; but a more general GLM is possible (with structure constants partly nilpotent) as example 3.5 makes clear. Some examples of super Lie groups will now be given.

<u>Example 3.3</u>

GL(4,4) (described in section 2) is a (32,32) dimensional super Lie group. Its sub groups include OSP(4,4), the group of supermatrices M satisfying

$$M^{ST}\begin{pmatrix} I & 0 \\ 0 & J \end{pmatrix} M = \begin{pmatrix} I & 0 \\ 0 & J \end{pmatrix} \qquad (3.4)$$

where I is the 4 × 4 identity

$$J = \begin{pmatrix} 0 & 1 & 0 & 0 \\ -1 & 0 & 0 & 0 \\ 0 & 0 & 0 & 1 \\ 0 & 0 & -1 & 0 \end{pmatrix}$$

and $\begin{pmatrix} A & X \\ Y & B \end{pmatrix}^{ST} = \begin{pmatrix} A^T & Y^T \\ -X^T & B^T \end{pmatrix}$

<u>Example 3.4</u>

Consider the set G of matrices of the form

$$\begin{pmatrix} 1 & x & y_1 \\ 0 & 1 & y_2 \\ 0 & 0 & 1 \end{pmatrix}$$

where x is an even element of B_L and y_1, y_2 are odd. The set G forms a group under matrix multiplication, and thus $E^{1,2}$ acquires the structure of a group G, the multiplication rule being summarised by

$$(x, y_1, y_2) \circ (u, v_1, v_2)$$
$$= (x + u, y_1 + v_1 + xv_2, y_2 + v_2). \qquad (3.5)$$

It is easy to see that G is a (1,2) dimensional super Lie group; its GLM can be characterised by three generators X, Y_1, Y_2 (X even, Y_1 and Y_2 odd) with graded commutators trivial apart from $[X_1, Y_2] = 2Y_1$.

Example 3.5

Here we give an example of a super Lie group whose GLM cannot be expressed as the tensor product of B_L and a graded Lie algebra. Consider $E^{1,1}$ with group operation

$$(x,y) \circ (u,v) = (x + u + ayv, y + v) \qquad (3.6)$$

where a is some even nilpotent element of B_L. The graded Lie module of H has generators {X,Y} with

$$[X,Y] = [Y,Y] = 0$$
$$[X,Y] = 2a\ X. \qquad (3.7)$$

In the graded manifold formalism a graded Lie group is defined to be a graded manifold (G,A) with G a conventional Lie group and A a sheaf of graded algebras over G with special properties. In particular one has a graded Lie algebra

$$\underline{g} = \underline{g}_0 \oplus \underline{g}_1 \qquad (3.8)$$

with $\underline{g}_0$ (the even part of $\underline{g}$) equal to the Lie algebra of G, and an extension to G of the representation of $\underline{g}_0$ on $\underline{g}_1$ defined by $\underline{g}$.

If one constructs the supermanifold corresponding to the graded manifold (G,A) by the process described in [4,5], the resulting supermanifold has a super Lie group structure with GLM equal to $B_L \otimes \underline{g}$ [13]; thus the correspondence between graded manifolds and a certain class of supermanifold can be extended to a correspondence between graded Lie groups and super Lie groups.

4 COMPACT SUPERMANIFOLDS WITH NON-ABELIAN FUNDAMENTAL GROUP

In the study of supermanifolds, two topologies have been considered on B_L (and $E^{r,s}$); one is the (finer) topology induced by the norm on B_L, and the other is the coarser non-Hausdorff topology introduced by De Witt [3]; in this coarser topology, U is open in B_L if and only if $U = \varepsilon^{-1}(V)$ for some V open in $\mathbb{R}$. (Also U is open in $E^{r,s}$ if and only if $U = \varepsilon_{r,s}^{-1}(V)$ for some V open in $\mathbb{R}^r$.) Up to this point, every supermanifold described would have been a supermanifold in the coarser topology (and thus a fortiori in the finer topology); in the fine topology such supermanifolds are not the most general possible kind, because they always have topologically the structure of a vector bundle over an underlying conventional manifold; the bundle will not in general be trivial, and so even using the coarse topology one has a source of genuinely non-trivial 'super-topology'. However there is a second possibility, only possible in the fine topology: the odd directions may be "curled up", so that the supermanifold is compact in some or all of the odd directions. Some examples of such supermanifolds are included in Batchelor's lectures [7]. A further example is given here, to show that the odd directions can be "curled up" in a topologically exotic manner.

Example 4.1

Consider the super Lie group G of example 3.4, that is $E^{1,2}$ with group action defined by equation (3.5). This group has a non-Abelian discrete subgroup D consisting of those elements (m,n_1,n_2) of $E^{1,2}$ where m, n_1, n_2 are "Grassmann integers", that is to say have integral coefficients when expanded in terms of some particular (non-canonical) basis of generators (and products of generators) of B_L. It can be shown [14] that the quotient space G/D is a (1,2) dimensional supermanifold which is compact. The usual exact sequence argument shows that the fundamental group of G/D is isomorphic to D.

5 DEVELOPMENTS AND APPLICATIONS

Clearly there are many aspects of supermanifold geometry which might profitably be examined; there are many theorems in the classical geometry of real and complex manifolds which might have analogues for supermanifolds. In particular there are theorems, such as de Rham's theorem and index theorems, which relate differentiable and topological structure. So far, these theorems have been considered in the graded manifold formalism (by Kostant [2] for de Rham's theorem and by Rempel and Schmitt [15] for super index theorems). These results are disappointing in that the truly 'super' aspects of the topology are not picked up; only the topology of the underlying manifold, and not the possible twistedness of the odd extension, features in the result. Obviously in the supermanifold formalism there is the possibility of extending these results to the finer topology.

At a less ambitious level, the question of integration (treated in a separate article [11]) is crucial, and far from clear; even with the coarser topology (which is easier because there are less open sets to handle) the consistency problem is not satisfactorily dealt with. Also, characteristic classes (and their integral representation), whose investigation was begun by Berezin [16], need further consideration.

One might also ask what applications there are for supermanifolds. Their application to supergravity has been mentioned in section 3 (and is described more fully in [9]); at the level of local coordinates they are a standard tool in supergravity theories. They might also be useful for an extension to supersymmetric theories of the 'topological' results for gauge theories; as an example, super index theorems might give a better understanding and control of the 'multiplet of anomalies' of supersymmetric theories, or a way to construct further supersymmetric models such as N = 2 d = 10 supergravity [17] with chiral fermions but no chiral anomaly [18].

Another area where supergeometric techniques are useful is lattice supersymmetry. The lattice approximation scheme involves approximating spacetime by a discrete lattice. The aim is to improve on conventional weak-field perturbation theory in two respects - first, handling situations such as strong coupling, where weak field perturbation theory is clearly inappropriate, and secondly enabling one to study analytically the approach to the true theory (that is to say, the limit, lattice spacing $\to 0$). Attempts have been made at putting supersymmetric theories on the lattice [19,20,21]; in addition to the motivation outlined above, there is also the hope that putting supersymmetry on the lattice might help with difficulty of putting fermions on the lattice. Since, in the supersymmetry algebra, the commutator of two supersymmetry transformations is a spacetime translation, one has two choices when putting supersymmetry on a lattice; either one must have discrete supersymmetry transformations, or one must sacrifice the supersymmetry algebra to some extent. Most authors do the latter; of relevance to this article are the approaches taken by Kostelecký and Rabin [20] and Scott [21]. Kostelecký and Rabin use a discrete superspace (that is to say, relative to the fine topology it is a discrete topological space) using "Grassmann integers" as in example 4.1. On a discrete subset of the real line one can allow a function to take arbitrary values, and still be able to interpolate with a C^∞ function, but one cannot let a function take arbitrary values on a discrete superspace and still interpolate with a G^{∞} function; Kostelecký and Rabin use a discrete version of the z-map of equation (1.4) to define the class of allowed functions on a superlattice, and proceed to translate the construction used on continuum superspace into lattice terms. Their approach is not entirely trouble free, because some of the constructions used lose their useful properties on the lattice, basically because finite difference operators do not obey a Leibnitz rule. In this approach, a discrete version of the Berezin integral is also required; on normal bosonic lattices one simply uses a sum (which tends to an integral in the continuum limit). Kostelecký and Rabin adopt the following approach: clearly a sum is desirable, because of the corresponding invariance, but in general the sums involved would appear to be infinite; they define the sum over an odd discrete variable to be symmetrised so that

$$\sum_{\theta} a\theta + b = Nb \tag{5.1}$$

(where N is the number of lattice points); but this symmetrisation gives then a well-defined sum for relevant cases at the expense of invariance under translations of the superlattice. This is only restored by the prescription (obviously closely related to Berezinian integration [16]) of differentiating before taking the sum.

An alternative approach to lattice supersymmetry is under investigation by Scott [21]; he considers a discrete subgroup not just of the translational but also of the rotational part of the super-Poincaré group. His proposal is to use the approach to lattice fermions in the Kähler formulation [22]. In this approach one begins with differential forms in the continuum, and then applies de Rham's theorem to translate onto the lattice; this has many appealing features, but there is a difficulty in that the Clifford product of differential forms (which is essential for the Kähler fermions) does not have a straightforward lattice analogue which is associative and respects the cubical symmetry of the lattice.

Thus no entirely satisfactory formulation of supersymmetry on the lattice has yet been found, but further consideration of the problem in a truly "supergeometric" spirit may resolve it.

REFERENCES

1. Berezin, F.A., and Leîtes, D.A., Sov. Math. Dokl. 16, 1218 (1975).

2. Kostant, B., "Graded manifolds, graded Lie theory and prequantization" in Differential Geometric Methods in Mathematical Physics, proceedings of the Symposium held at Bonn, July 1975, Lecture Notes in Mathematics 570 (Springer Verlag, 1977).

3. De Witt, B.S., Austin University Preprint (1977), (to be published).

4. Batchelor, M., Trans Am. Math. Soc. 258, 257 (1980).

5. Rogers, A., Journ. Math. Phys. 21 1352 (1980).

6. Jadczyk, A., and Pilch, K., Comm. Math. Phys. 78, 373 (1981).

7. Batchelor, M., lectures in in this volume, p. 91.

8. Volkov, D.V., and Akulov, V.P., Phys. Lett. 46B, 109 (1973). Abdus Salam and Strathdee, J., Nuc. Phys. B76, 477 (1974).

9. Wess, J., and Zumino, B., Phys. Lett. 66B, 361 (1977).

10. Bonora, L., Pasti, P., and Tonin, M., Journ. Math. Phys. 23 839 (1982).

11. Rogers, A., "Integration on Supermanifolds", article in this volume, p. 135.

12. Rogers, A., "The geometry of curved superspace and supergravity", London University Ph.D. Thesis (1981).

13. Rogers, A., Journ. Math. Phys. 22 939 (1981).

14. Rogers, A., Journ. Math. Phys. 22 443 (1981).

15. Rempel, S., and Schmitt, T., Seminar. Analysis, 1981/82, 92-131, Akad. Wiss. DDR, Berlin 1982.

16. Berezin, F.A., Sov. J. Nucl. Phys. 30, 605 (1979).

17. Nahm, W., Nucl. Phys. B135 (1978) 149.
Green, M.B., and Schwarz J.H., Phys. Lett. 109B, 444 (1982), 122B 143 (1983).
Schwarz, J.H., and West, P., Phys. Lett. 126B, 301 (1983).
Howe, P.S., and West P., King's College preprint, (1983)

(to be published in Nucl. Phys. B).
Schwarz, J.H., 1983 Caltech preprint.

18. Alvarez-Gaumé, L., and Witten E., 1983 Princeton Preprint.

19. Dondi, P.H., and Nicolai, H., Nuov. Cim. 41A (1977) 1.
Banks, T., and Windey, P., Nucl. Phys. B198 (1982), 226.
Rittenberg, V. and Yankielowicz, S., CERN TH 3263 (1982).
Elitzur E., Rabinovici, E., and Schwimmer, A., Phys. Lett. 119B 165 (1982).

20. Kostelecký, V.A., and Rabin, J.M., Los Alamos preprint LA-UR-83-1373 (1983).

21. Scott, D., Edinburgh University Preprint 83/255 (1983).

22. Becher, P., and Joos, H., Desy Preprint 82-031 (1982).

INTEGRATION ON SUPERMANIFOLDS

Alice Rogers

The Blackett Laboratory
Imperial College, London SW7 2BZ

Research Supported by the SERC

Abstract The theory of integration on supermanifolds, based on the work of Berezin [1] , is described. A summary of classical integration theory is given, followed by an explanation of the inapplicability to supermanifolds of a straightforward generalisation of classical integration theory. The basic rule for integration over odd Grassmann variables is given. Superforms are defined, and the rule for integrating a superform over a supermanifold is given. An exterior derivative on superforms is defined, and the corresponding Stokes' theorem discussed.

1. INTRODUCTION

The aim of this lecture is to show how a consistent theory of integration may be developed on supermanifolds, using the approach due to Berezin [1] . The physical motivation for studying integration over odd as well as even variables includes the need for functional integrals over fermion fields and the construction of invariants in supersymmetric quantum field theories. Integration theory over odd variables is so strange that one might expect it to be simply a mathematical curiosity, whereas it is in fact a valuable tool regularly used by theoretical physicists. As is indicated in more detail in §5 and 6, the mathematical theory is far from complete.

2. STANDARD INTEGRATION THEORY

We begin with a brief statement, following Hicks [2] , of standard integration theory on conventional, real manifolds. Here the fundamental definition is the integral of a p-form over a p-chain on a manifold M : suppose $\sigma = \sum r_i \sigma_i$ is a real p-chain on M, with each σ_i a p-simplex on M, that is to say

H.-J. Seifert et al. (eds.), Mathematical Aspects of Superspace, 149–160.

a C^∞ map from the unit p-cube I^p into M, and the coefficient r_i real numbers. Also suppose that w is a p-form on M. Then by definition

$$\int_\sigma w := \sum r_i \int_{I^p} \sigma_i^* w \tag{2.1}$$

Here each $\sigma_i^* w$ is a p-form on I^p and can be evaluated as a Riemann integral in the usual way. More "practically" one wishes to integrate a function over a manifold M, or a subset U of M; in this case M must be an orientable Riemannian (or pseudo-Riemannian) manifold, so that there is a volume form, and M (or U) must be built up by patching together the images of p-cubes for some value of p. In more detail, one requires a 'fundamental p-chain', that is to say a p-chain $\sigma = \sum r_i \sigma_i$ with each r_i equal to one, each σ_i being injective and orientation preserving and the images of the σ_i only coinciding on their boundaries; defining the support of σ to be $\bigcup_i \operatorname{Im} \sigma_i$, the final patching requirement is that the support of σ should equal M(or U). One then defines

$$\int_M f := \int_\sigma f v. \tag{2.2}$$

where v denotes the volume form on M.

Two questions clearly arise,

(a) is this patching possible and

(b) if two different patchings of M are possible, do they give the same value for the integral of any f over M?

The answer to (a) is that the patching is always possible if M is compact [3] , but is not always possible in more general cases. The answer to (b) is that two different patchings will give the same answer; a proof due to King Lee is quoted in Hicks [2, p.103] . The result ultimately depends on two properties of Riemannian integration on $\mathbb{R}^n$,

(i) the additive property, given X,Y in R with $\operatorname{int} X \cap \operatorname{int} Y = \phi$, then

$$\int_X f\, d^n x + \int_Y f\, d^n x = \int_{X \cup Y} f\, d^n x \tag{2.3}$$

and (ii) the rule for change of variables by the Jacobian,

$$\int_{f(U)} w = \int_U f^* w \tag{2.4}$$

where $f = U \to \mathbb{R}^n$ is a diffeomorphism of U on to a subset of $\mathbb{R}$. The proof of this result, which is not trivial, may be found in Whitney [4].

While the patching technique extends the "limit of a sum" aspect of elementary integration to manifolds, there is also Stokes' theorem

$$\int_{\sigma} d\omega = \int_{\partial\sigma} \omega \tag{2.5}$$

which extends the "differentiation backwards" aspect of elementary integration to manifolds.

3. INTEGRATION OVER ODD VARIABLES

Turning now to supermanifolds, we wish to build up a similar theory, starting from integrals over the flat "super Euclidean" space $E^{r,s}$ with r even Grassmann co-ordinates and s odd Grassmann co-ordinates and then patching to obtain integrals over more general supermanifolds. (Detailed descriptions of $E^{r,s}$ and supermanifolds may be found in Batchelor's article in this volume [5]). While some differential geometry on supermanifolds (such as tensor calculus) is a relatively predictable generalisation of conventional differential geometry, this is certainly not the case for integration.

It is obvious that integration in the sense of "differentiation backwards" will pose problems when dealing with odd variables, since (considering just one odd variable y) one cannot even find a function whose derivative is the simple function $f : y \to y$. Another difficulty is that the natural choice of differential form on a supermanifold, that is to say graded-antisymmetric covariant tensor fields, as described by Kostant [6] , do not have an upper limit on their degree. This is most easily seen in a local co-ordinate system; in a co-ordinate system $(x_1,\ldots,x_r;\, y_1,\ldots,y_s)$ on an (r, s) dimensional supermanifold, a differential form consists of terms such as

$$f(x,y)\,dx^{p_1}\wedge\ldots\wedge dx^{p_i}\wedge dy^{q_1}\wedge\ldots\wedge dy^{q_j}\,;$$

the dx^{p_k} anticommute with one another and so there cannot be more than r such factors, but the dy^{q_k} commute with one another and so are unrestricted.

Related to the lack of 'top forms' on supermanifolds are the properties of determinants of so-called "supermatrices". An (r,s) x (r,s) supermatrix is a square matrix M with block diagonal form

$$M = \begin{pmatrix} A & Y \\ X & B \end{pmatrix} \tag{3.1}$$

where A is r x r, B is s x s and A and B have even Grassmann entries, while X, Y have odd Grassmann entries. Such matrices define transformations of $E^{r,s}$ which are the natural analogue of the usual linear transformations of Euclidean space; also, of particular relevance to the subject of integration, the Jacobian matrix

$$\left(\partial z^k / \partial z'^l \right) = \begin{pmatrix} \partial x^i / \partial x'^j & \partial x^i / \partial y'^n \\ \partial y^m / \partial x'^j & \partial y^m / \partial y'^n \end{pmatrix} \tag{3.2}$$

is of this form. Normally with two matrices M and N one has the rule

trace MN = trace NM .

However, this property is not true in general for supermatrices, as a simple counter example with (1,1) x (1,1) supermatrices demonstrates :

$$\begin{aligned} \operatorname{tr} \begin{pmatrix} a & x \\ y & b \end{pmatrix} \begin{pmatrix} c & u \\ v & d \end{pmatrix} &= ac + xv + yu + bd \\ \operatorname{tr} \begin{pmatrix} c & u \\ v & d \end{pmatrix} \begin{pmatrix} a & x \\ y & b \end{pmatrix} &= ac - xv - yu + bd \end{aligned} \tag{3.3}$$

However, if a new 'supertrace' [7,8,9] is defined by

$$\operatorname{str} \begin{pmatrix} A & X \\ Y & B \end{pmatrix} = \operatorname{tr} A - \operatorname{tr} B \tag{3.4}$$

one regains the law

$$\operatorname{str} MN = \operatorname{str} NM . \tag{3.5}$$

At the level of determinants, the corresponding difference is that, if one uses a naive generalisation of the usual determinant, then one loses the vital property

$$\det A \ \det B = \det AB . \tag{3.6}$$

The modified definition of determinant which restores this vital property can be deduced from the rule Ber M = exp (str ln M) where Ber (short for Berezinian) denotes the super determinant [7,8,9] .

The result is that if

$$M = \begin{pmatrix} A & X \\ Y & B \end{pmatrix} \tag{3.7}$$

and $$M^{-1} = \begin{pmatrix} C & U \\ V & D \end{pmatrix}$$

then Ber M = det A det D

or equivalently

$$\mathrm{Ber}\ M = \det(A - XB^{-1}Y)\ \det B^{-1} \tag{3.8}$$

if B has no inverse the determinant is undefined. A full treatment of the superdeterminant is given in [9] and [10] .

Having mentioned some of the unexpected features of Grassmann integration, we can now see how it may be carried out; one loses both the "limit of a sum" and the "differentiation backwards" aspect. The starting point is the well known rule [8] for "Berezin Integration" over an odd Grassmann variable y :

$$\int y \text{ "}dy\text{"} = 1 \qquad\qquad \int 1 \text{ "}dy\text{"} = 0 \tag{3.9}$$

By linearity one can now differentiate over the most general "superdifferentiable" function of y. (Various definitions of superdifferentiable, all equivalent in this situation, are discussed in [5]). It is an open question whether the concept of integration can be extended to more general functions. It should be noted that there are no limits of integration, in other words one is using the coarser De Witt topology (again refer to [5]) with its "all or nothing" approach to the odd part of the Grassmann algebra.

The motivation for the introduction of this rule is that physical applications require translational invariance. (Various authors give reasons for this, for instance Van Nieuwenhuizen [10, appendix] and Wilson [11]). It is the only rule of integration which gives translational invariance together with non-trivial but finite results for the (very restricted) class of functions being considered. The only non-trivial multiple integral over s odd variables $y_1, \ldots, y_s$ is defined with the sign convention

$$\int y_1 y_2 \cdots y_n \, dy_n \cdots dy_2 \, dy_1 \tag{3.10}$$

We now wish to see whether the various properties of classical integration theory which are used to build up a theory of integration on conventional manifolds have their analogues for supergeometry. A crucial property is good functorial behaviour of integrals, and for this one needs to determine the nature of the "differential" which is denoted "dy" in equation (3.9). Proceeding heuristically, consider how $\int y$ "dy" changes if we reparametrise the range of integration by $y' = a\,y$, where a is an invertible, even Grassmann element. The 'range' of integration is preserved since we are integrating over the entire odd Grassmann algebra ($E^{0,1}$ in the notation of [5]). Thus to preserve $\int y$ "dy" = 1 we require "dy′" = $\frac{1}{a}$ "dy"; this is of course the reverse of the usual situation, and reveals immediately that the "dy" are not simply odd co-ordinate differentials.

4. SUPERFORMS

The heuristic discussion of the preceding section shows that a new type of form, with a contravariant transformation rule, is required for integration over odd variables. One of course continues to need the usual forms for even integration, so that integration on supermanifolds requires a hybrid object, called a superform. These were first defined by Berezin [1] ; a different but equivalent definition, applicable to all the various categories of supermanifolds and graded manifolds (cf. Batchelor's article in this volume [5]) will be given here. The definition depends ultimately on the algebra of smooth functions on the supermanifold (or graded manifold) in question; following Batchelor [5] we will use S(U) to stand for the set of superdifferentiable (in the appropriate sense) functions on an open set U of the supermanifold (or an open set U of the underlying manifold of a graded manifold).

Definition 4.1 Let $D^1(U)$ and $D_1(U)$ denote respectively the S(U) modules of vector fields and one-forms on U. A (p,q) superform on U is a graded S(U)-linear map

$$\alpha:\ D^1(U)^p \otimes D_1(U)^q \to S(U)$$
$$(X_1,\dots,X_p;\omega_1,\dots,\omega_q) \mapsto \langle\alpha|X_1,\dots,X_p;\omega_1,\dots,\omega_q\rangle$$

such that

$$\langle\alpha|X_1,\dots,X_i,X_{i+1},\dots,X_p;\omega_1,\dots,\omega_q\rangle$$
$$=(-1)^{1+|X_i||X_{i+1}|}\langle\alpha|X_1,\dots,X_{i+1},X_i,\dots,X_p;\omega_1,\dots,\omega_q\rangle$$

and

$$\langle\alpha|X_1,\dots,X_p;\omega_1,\dots,\omega_i,\omega_{i+1},\dots,\omega_q\rangle$$
$$=(-1)^{|\omega_i||\omega_{i+1}|}\langle\alpha|X_1,\dots,X_p;\omega_1,\dots,\omega_{i+1},\omega_i,\dots,\omega_q\rangle \qquad (4.1)$$

The exterior product of such forms can be defined in the expected manner, and in a local co-ordinate system z^i (i=1,...,m+n) on an (m,n) dimensional supermanifold has an expansion in terms of the basic co-ordinate differentials dz^i and derivatives $\frac{\partial}{\partial z^i}$ of the form

$$\alpha = \sum_{i_k, j_k = 1}^{m+n} f^{i_1 \dots i_p j_1 \dots j_q} dz^{i_1} \wedge \dots \wedge dz^{i_p} \wedge \frac{\partial}{\partial z^{j_1}} \dots \wedge \frac{\partial}{\partial z^{j_q}} \qquad (4.2)$$

Denoting $z^i = x^i \; (1 \leq i \leq m)$ and $z^i = y^i \; (1+m \leq j \leq n+m)$

one finds $dx^{i_1} \wedge dx^{i_2} = - dx^{i_2} \wedge dx^{i_1}$

and

$$\frac{\partial}{\partial y^{j_1}} \wedge \frac{\partial}{\partial y^{j_2}} = - \frac{\partial}{\partial y^{j_2}} \wedge \frac{\partial}{\partial y^{j_1}} \qquad (4.3)$$

This means that there does exist a quasi "top-form" in that an (m,n) form on an (m,n) supermanifold will split, in a given set of local co-ordinates, into

$$\alpha = \alpha_0 + \alpha_1$$

where

$$\alpha_0 = f(x,y)\, dx^1 \wedge \dots \wedge dx^m \wedge \frac{\partial}{\partial y^1} \wedge \dots \wedge \frac{\partial}{\partial y^n} \qquad (4.4)$$

while all the terms in the expansion of α_1 contain either some dy^j or some $\frac{\partial}{\partial x^i}$ (or possibly both). Under change of co-ordinates $z \to z'$ the "volume element" $dx^1 \wedge \dots \wedge dx^m \wedge \frac{\partial}{\partial y^1} \wedge \dots \wedge \frac{\partial}{\partial y^n}$ changes by a factor equal to the Berezinian superdeterminant of the Jacobi matrix $(\partial z'^k / \partial z^i)$.
Berezin defines an <u>integral</u> superform to be a superform which has exactly

$$\alpha = \alpha_0 \qquad (4.5)$$

in some co-ordinate system; (of course, on transformation to another co-ordinate system, α_1 type terms may appear, but the leading term will transform according to the Berezinian superdeterminant).

An exterior derivative can be defined on superforms, but it is a trivial extension of the conventional exterior derivative in that it only affects the covariant part of the superform :

Given

$$\alpha = \sum_{i_k, j_k = 1}^{m+n} f^{i_1 \dots i_p j_1 \dots j_q} dz^{i_1} \wedge \dots \wedge dz^{i_p} \wedge \frac{\partial}{\partial z^{j_1}} \wedge \dots \wedge \frac{\partial}{\partial z^{j_q}} \tag{4.6}$$

$$d\alpha := \sum_{i_k, j_k = 1}^{m+n} \sum_{l=1}^{m+n} \frac{\partial f}{\partial z^l} dz^l \wedge dz^{i_1} \wedge \dots \wedge dz^{i_p} \wedge \frac{\partial}{\partial z^{j_1}} \wedge \dots \wedge \frac{\partial}{\partial z^{j_q}}.$$

5. INTEGRATION ON $E^{r,s}$.

Using the coarse topology on $E^{r,s}$ it is now possible to give a meaning to $\int \alpha$ where α is an integral (r,s) superform [1]. By definition, if α is an integral (m, n) superform on a open subset U of an (m, n) dimensional supermanifold, and $\emptyset$ is a co-ordinate map such that in the corresponding local co-ordinates z = (x, y)

$$\alpha = f(x,y)\, dx^1 \wedge \dots \wedge dx^m \wedge \frac{\partial}{\partial y^1} \wedge \dots \wedge \frac{\partial}{\partial y^n}$$

then

$$\int_U \alpha = \int_{\varepsilon \circ \phi(U)} f \circ \phi^{-1}(u,v)\, d^m u\, d^n v \tag{5.1}$$

where integration over the odd variables is carried out by the rule (3.9) and integration over the even variables by the usual methods. Here ε denotes the augmentation map projecting $E^{m,n}$ onto $\mathbb{R}^m$. (In the graded manifold formalism the ε is not required). In other co-ordinate systems, where α may have additional terms, containing dy or $\partial/\partial x$, the rule is simply to ignore these terms and consider only the leading term.

Since our aim is to integrate over supermanifolds, employing some patching scheme, it is clearly important to consider the two properties an integral must have in order for such patching to be possible,

(i) the additive property and

(ii) the change of variable property

(described in § 2).

Using the coarse topology, one has the additive property

$$\int_U \alpha + \int_V \alpha = \int_{U \cup V} \alpha \tag{5.2}$$

precisely as in the classical case, and as a direct consequence of that result. The consistency under change of variable property (ii) is only partly retained, a fact which does not seem to have been recognised in the literature. What one wishes is, given U open in $E^{m,n}$,

$$\int_U f(x)\, d^m x\, d^n y = \int_{g(U)} f \circ g^{-1}(x)\, \mathrm{Ber}(g)\, d^m x\, d^n y \qquad (5.3)$$

where g is any invertible map of U into $E^{m,n}$.

Obviously if n = 0, one is simply considering the well known result from classical integration theory. At the other extreme, if m = 0, it is relatively simple to see that the result holds, simply because of the anticommutativity property of the odd co-ordinates. It is the case where odd and even coexist and intermingle that leads to difficulties. If one expresses the transformation g in terms of local co-ordinates as g (x,y) = (x',y'), one may break down g into two stages, first changing y but leaving x unchanged, and then changing x while holding y' fixed. An ingenious proof due to Fung and quoted in the appendix of [10] shows that the result (5.3) holds for the first stage of g. But it is easy to construct a counter example to show that (5.3) does not hold for the second type of transformation.

Consider $E^{1,1}$ and the integral

$$\int_{\varepsilon^{-1}((0,1))} x\, dx\, dy$$

By the Berezin rule this integral gives zero. However, if one makes the transformation

$$\begin{aligned} g &: \varepsilon^{-1}((0,1)) \to E^{1,1} \\ &(x,y) \rightsquigarrow (x+\eta y, y) \end{aligned} \qquad (5.4)$$

where η is a constant odd Grassmann element one finds Ber (g) = 1 and thus the transformed integral is

$$\int_{\varepsilon^{-1}((0,1))} (x+\eta y)\, dx\, dy = \eta\,. \qquad (5.5)$$

Thus one finds that the integration theory developed lacks the complete consistency that one might desire; even with the more restricted set of smooth functions used by Kostant [6] and Berezin and Leites [12], examples can easily be constructed to show that consistency cannot be maintained. A proof by De Witt [9] shows that for infinitesimal transformations the error in (5.3) is a surface integral; this result is sufficient for physical applications so far considered.

Finally, it should be pointed out that corresponding to the exterior derivative (4.6) there is an obvious but somewhat trivial Stokes' theorem; trivial in that the dy terms are thrown away and the boundary is simple the boundary in the underlying real manifold.

6. INTEGRATION ON SUPERMANIFOLDS

In order to integrate a function over a supermanifold one needs the same two features as in the classical theory, a volume form and a patching together of the supermanifold by manageable pieces. The volume form can be derived from a metric in the usual way; most applications are in supergravity and here the volume form (analogous to det $e_m{}^a\, d^4x$ in classical gravity) is Ber $E_m{}^A\, d^m x\, d^n y$ where $(E_M{}^A)$ is the 'Vielbein' matrix (c.f. Wess [13]). If one uses a graded manifold, or a supermanifold modelled on $E^{r,s}$, with the coarse topology, it is possible to patch together a supermanifold simply by patching together the underlying manifold as in § 2. But clearly here the consistency problem rears its head, as a consistent definition can only be given on supermanifolds which admit an atlas whose transition functions do not violate (5.3). The manifolds used in supergravity all have this property (c.f. [16]) but it seems that consistency may restrict one to trivial supermanifolds.

This completes the description of integration theory on supermanifolds as it stands at present; in its somewhat incomplete form it is, as has already been stated, regularly used by physicists. There are, of course, many features of the theory which have yet to be explored. First of all there is the question, what exactly _is_ Berezin integration? An algebraic geometric answer to this question is given by Batchelor [14] . Here the point of view is taken that Berezin integration not only looks like differentiation but actually is differentiation, and that the superform is a combined form and chain; both Berezin integration and Riemannian integration are seen as particular cases of a much wider formalism.

Obviously the question "what is Berezin integration" can become mere semantics, but it would be desirable to find some analogue of the "differentiation backwards" or the "limit of a sum" aspect of conventional integration. The obvious analogue of the "differentiation backwards" aspect would be a Stokes' theorem involving a more interesting exterior derivative than that of (4.6), one that actually involved the $\partial/\partial z$ as well as the dz. A heuristic Stokes' law (relying on the fact that any function which is an odd derivative must have one y factor missing and thus integrate to zero) is regularly used by physicists; but this has not been expressed formally in terms of an exterior derivative on superforms.

The "limit of a sum" aspect of integration is clearly lost; indeed a naive attempt at integration over 'bits' of the odd part of the Grassmann algebra (that is to say, using the fine topology) clearly fails. A simple example reveals the problem : working with a Grassmann algebra with one generator β , following the usual additive principle one might expect to find

$$\int_0^{\beta} y\,dy + \int_{\beta}^{2\beta} y\,dy = \int_0^{2\beta} y\,dy \tag{6.1}$$

while in fact, because of the transformation rule for superforms one finds that

$$\int_0^{\beta} y\,dy = \int_{\beta}^{2\beta} y\,dy \tag{6.2}$$

but

$$\int_0^{2\beta} y\,dy = \int_0^{\beta} y\,dy \tag{6.3}$$

and so (6.1) cannot hold.

However, even if one cannot retain the additive property, one certainly wishes integration to take account of the value of a function "everywhere" and not just at a point; in this sense Berezin integration resembles contour integration, where certain integrals can be evaluated by looking at a function near a single point, but do really take into account the value of the function over the full range of integration, the latter being determined by the former by analytic continuation.

To summarise, outstanding problems in supermanifold integration theory include the lack of full consistency under change of co-ordinates, the need for a complete Stokes' theorem and the extension of the theory to enable one to integrate over pieces of the odd Grassmann variable. This is related to the problem of defining supersymmetric theories on a lattice [15] and is described more fully in the accompanying article [16] .

REFERENCES

1. Berezin, F.A. 1979, Sov.Journ.Nucl.Phys. 30 p.605.

2. Hicks, N.J. 1965, "Notes on Differential Geometry", Van Nostrand, New York.

3. Cairns, S.C. 1961, Bull.Am.Math.Soc. 67, p.389.

4. Whitney, H. 1957, "Geometric Integration Theory", Princeton.

5. Batchelor, M. 1984, article in this volume, p. 91.

6. Kostant, B. 1977, "Graded manifolds, graded Lie theory and prequantization" in Differential Geometric Methods in Mathematical Physics, Lecture notes in physics 570, Springer, pp 177-306.

7. Arnowitt, R. Nath, P. and Zumino, B. 1975, Physics Letters B56, p. 82.

8. Berezin, F.A., 1966, "The Method of Second Quantization," Academic Press, New York.

9. De Witt, B.S. 1977, Austin University Preprint (to be published).

10. Van Nieuwenhuizen, P., 1981, Physics Reports 68, p. 189.

11. Wilson, K.G., 1974, Phys.Rev. D10, p. 2445.

12. Berezin, F.A. and Leites, D. 1975, Sov.Math.Dokl. 16, p. 1218.

13. Wess, J., article in this volume, p. 1.

14. Batchelor, M., 1982, "Algebraic Integration," Cambridge University preprint.

15. Dondi, P.H. and Nicolai, H. 1977, Nuovo Cimento 41A, p.1.
Banks, T. and Windey, P., 1982, Nucl.Phys. B198, 226.
Rittenberg, V. and Yankielowicz, S., 1982, CERN preprint TH3263,
Elitzur, E., Rabinovici, E., and Schwimmer, A., 1982 Phys. Lett. 119B, p.165.
Scott, D.M., 1983, Edinburgh University preprint no. 83/255.

16. Rogers, A., article in this volume, p. 135.

REMARKS ON BATCHELOR'S THEOREM

Robert J. Blattner* & John H. Rawnsley

Department of Mathematics, University of California, Los Angeles, CA 90024

Mathematics Institute, University of Warwick, Coventry, CV4 7AL

We show that the structure sheaf of a graded manifold is the sheaf of sections of the exterior algebra bundle of a vector bundle by reducing the problem to one in the theory of extensions of sheaves of commutative algebras to which standard cohomological techniques can be applied.

In [4] B. Kostant described a mathematical theory of graded manifolds. The basic objects consist of a manifold X and a sheaf A of $\mathbb{Z}_2$-graded commutative algebras over X with an epimorphism

$$A \xrightarrow{\varepsilon} C$$

of sheaves of $\mathbb{Z}_2$-graded algebras (where $(C^\infty)_0 = C^\infty$, $(C^\infty)_1 = 0$) which is locally isomorphic to

$$C^\infty|_U \otimes \Lambda \mathbb{R}^s \longrightarrow C^\infty|_U$$

in some neighbourhood U of each point of X. M. Batchelor [1] showed that there is a vector bundle E over X such that A is isomorphic to the sheaf of sections of the exterior algebra bundle ΛE. In her lectures in this volume Batchelor gives a proof using coalgebras. P. Green [3] has shown that if C^∞ is replaced by the sheaf of germs of holomorphic functions (for X a complex

*Research partially supported by NSF grant MCS 78-01332

H.-J. Seifert et al. (eds.), Mathematical Aspects of Superspace, 161–171.

manifold), then Batchelor's result is no longer true. We shall give a proof of Batchelor's Theorem using sheaf-theoretical methods which work in a very general setting. It is not the shortest proof possible (M. Rothstein [5] has a much more direct proof) but the generality of our method makes it available in other situations besides that of smooth graded manifolds. (The very short proof in [2] is incorrect.)

Instead of $\Lambda\mathbb{R}^s$ as the local model we shall assume we have a field k of characteristic 0, a finite dimensional algebra D with 1 over k, and a homomorphism

$$\varepsilon_{\circ}: D \longrightarrow k$$

of algebras with $\varepsilon_{\circ}(1)=1$, $J=\operatorname{Ker}\varepsilon_{\circ}$ a nilpotent ideal ($J^{n+1}=0$ for some $n \geq 1$). Moreover, we shall replace C^{∞} by an arbitrary sheaf C of commutative k-algebras with 1. (We make the convention that all algebras have 1 and all algebra morphisms send 1 to 1).

Let X be a topological space and C a sheaf of commutative k-algebras and $(D, \varepsilon_{\circ})$ be as above. By a <u>D-manifold structure over</u> X we shall mean a sheaf A of algebras on X with a morphism

$$\varepsilon : A \longrightarrow C$$

of sheaves of algebras such that X has a covering $\{U_\alpha\}$ by open sets with isomorphisms

$$\tau_\alpha : A|_{U_\alpha} \qquad C|_{U_\alpha} \otimes D$$

with

$$(\mathrm{id} \otimes \varepsilon_{\circ}) \circ \tau_\alpha = \varepsilon .$$

Obviously, taking $k=\mathbb{R}$, X a manifold, $C=C^{\infty}$ and $D=\Lambda\mathbb{R}^s$ we recover Kostant's graded manifolds, provided one takes into proper account the Z_2-grading.

By a <u>vector bundle over</u> X <u>of class</u> C we shall mean a locally free sheaf E of C-modules. Thus X can be covered by open sets U_α on which we have isomorphisms

$$\sigma_\alpha : E|_{U_\alpha} \longrightarrow C|_{U_\alpha} \oplus \ldots \oplus C|_{U_\alpha} = C|_{U_\alpha} \otimes k^r$$

of sheaves of C-modules. The transition functions of E are given by

$$g_{\alpha\beta} = \sigma_\alpha \circ \sigma_\beta^{-1}$$

and these are rxr matrices with entries in C. It is not necessary to have the vector space k^r in the definition of a vector bundle. If D is any finite dimensional k vector space and if we have local isomorphisms

$$\tau_\alpha : E|_{U_\alpha} \longrightarrow C|_{U_\alpha} \otimes D$$

which are C-module maps, then E is locally free (choose a basis for D) and E has structure group GL(D).

If D is an algebra and the transition functions are automorphisms of D (more precisely C-linear automorphisms of $C \otimes D$), then we say E has structure group Aut(D). Then E can be given the structure of a sheaf of algebras so that each τ_α is an isomorphism of sheaves of algebras. If additionally we have a homomorphism

$$\varepsilon_0 : D \longrightarrow k$$

and each transition function preserves $id \otimes \varepsilon_0$ we say E has structure group Aut(D, ε_0). Then E has a homomorphism

$$\varepsilon : E \longrightarrow C$$

such that locally

$$(id \otimes \varepsilon_0) \circ \tau_\alpha = \varepsilon.$$

Thus if we have a vector bundle of class C with structure group Aut(D, ε_0) then it is a D-manifold structure over X. Our purpose is to study the obstrutions to any given D-manifold structure A being such a vector bundle.

Now a D-manifold structure A over X is not necessarily a vector bundle since we did not assume that A had any C-module structure. Thus to see that a given D-manifold stucture A over X <u>is</u> a vector bundle we have to find a C-module structure on A and show it is locally free. In the case ultimately of interest where $D = \Lambda R^s$ and $A = \Lambda E$ we have a homomorphism $: C = \Lambda^0 E \hookrightarrow Z(\Lambda E)$ which splits ε, where Z(D) denotes the centre of the algebra D and Z(A) denotes the sheaf whose stalks are the centres of the stalks of A (this notation will be applied to other algebras and sheaves of algebras). Such a morphism is sufficient in the general case.

Proposition 1. Let $\varepsilon : A \longrightarrow C$ be a D-manifold structure over X, and suppose that $\varphi : C \longrightarrow Z(A)$ is a morphism of sheaves of algebras such that $\varepsilon \circ \varphi = \mathrm{id}_C$. Then A is locally free as a sheaf of C-modules and has structure group $\mathrm{Aut}(D, \varepsilon_0)$.

To prove this we shall need the following Lemma, whose proof is standard and is omitted:

Lemma 1. Let D be an algebra over a field k of characteristic 0 and θ a unipotent endomorphism of D. Then

$$\log\theta = \sum_{p=1}^{\infty} (-1)^{p-1} (\theta - \mathrm{id})^p/p \qquad \text{(a finite sum)}$$

is a nilpotent derivation of D. If δ is a nilpotent derivation of D then

$$\exp\delta = \mathrm{id} + \sum_{p=1}^{\infty} \delta^p/p! \qquad \text{(a finite sum)}$$

is a unipotent endomorphism of D. log and exp are inverses of each other and give a bijection between the set of unipotent automorphisms of D and the set of nilpotent derivations of D.

Proof of Proposition 1: Define an endomorphism θ_α of $C|_{U_\alpha} \otimes D$ by

$$\theta_\alpha(c \otimes d) = \tau_\alpha(\varphi(c)) 1 \otimes d.$$

Since $\varphi(c) \in Z(A)|_{U_\alpha}$, $\tau_\alpha(\varphi(c)) \in C|_{U_\alpha} \otimes Z(D)$, θ_α is indeed an endomorphism of algebras, and

$$\theta_\alpha(c \otimes d) - c \otimes d = [\tau_\alpha(\varphi(c)) - c \otimes 1] 1 \otimes d.$$

But

$$\mathrm{id} \otimes \varepsilon_0 (\tau_\alpha(\varphi(c)) - c \otimes 1) = \varepsilon(\varphi(c)) - c = 0,$$

so $\tau_\alpha(\varphi(c)) - c \otimes 1 \in C|_{U_\alpha} \otimes J$. It follows that $\theta_\alpha - \mathrm{id}$ maps $C|_{U_\alpha} \otimes J^p$ into $C|_{U_\alpha} \otimes J^{p+1}$, so $(\theta_\alpha - \mathrm{id})^{n+1} = 0$. Thus θ_α is unipotent and so, by Lemma 1, is of the form $\exp\delta$ and hence invertible. Thus

$$\sigma_\alpha = \theta_\alpha^{-1} \circ \tau_\alpha : A|_{U_\alpha} \longrightarrow C|_{U_\alpha} \otimes D$$

is an isomorphism of sheaves of algebras, and

$$\sigma_\alpha(\varphi(c)a) = \theta_\alpha^{-1}(\tau_\alpha(\varphi(c)))\theta_\alpha^{-1}(\tau_\alpha(c))$$

$$= c\otimes 1\,\sigma_\alpha(a).$$

Thus σ_α is also an isomorphism of C-modules. Hence A is locally free, with transition functions $\sigma_\alpha \cdot \sigma_\beta^{-1}$ which have values in Aut(D). But

$$\begin{aligned} id\otimes\varepsilon_0(\theta_\alpha(c\otimes d)) &= id\otimes\varepsilon_0(\tau_\alpha(\varphi(c))1\otimes d) \\ &= \varepsilon\varphi(c)\,\varepsilon_0(d) \\ &= c\,\varepsilon_0(d) \\ &= id\otimes\varepsilon_0(c\otimes d). \end{aligned}$$

Thus θ_α and hence θ_α^{-1} preserve $id\otimes\varepsilon_0$ so that $\sigma_\alpha \cdot \sigma_\beta^{-1}$ has values in Aut(D, ε_0) as claimed. □

As a consequence of Proposition 1 our problem has reduced to studying when

$$\varepsilon : Z(A) \longrightarrow C$$

is split as an epimorphism of sheaves of algebras.

We let $J_0 = Z(D)\cap J$, so we have the restricted epimorphism

$$\varepsilon_0 : Z(D) \longrightarrow k$$

with kernel J_0. Of course J_0 is also nilpotent. If $I = \mathrm{Ker}\,\varepsilon$ then τ_α maps $I|_{U_\alpha}$ isomorphically onto $C|_{U_\alpha}\otimes J$, and $I^p|_{U_\alpha}$ onto $C|_{U_\alpha}\otimes J^p$. Moreover τ_α maps $Z(A)|_{U_\alpha}$ to $C|_{U_\alpha}\otimes Z(D)$ and if $I_0 = I\cap Z(A)$, then I_0 is mapped to $C|_{U_\alpha}\otimes J_0$ by τ_α. Thus

$$\varepsilon : Z(A) \longrightarrow C$$

is a Z(D)-manifold structure over X. This construction and Proposition 1 allow us to reduce our problem to splitting $\varepsilon : A \longrightarrow C$ where A is commutative.

We assume until further notice that D is commutative.

Let $\varepsilon : A \longrightarrow C$ be a D-manifold structure over X and let I and J be as above. Then we have an induced map $\varepsilon_p : A/I^p \longrightarrow C$ with kernel I^p which makes A/I^p a D/J^p-manifold structure over X. Multiplication by elements of A/I^p on I^{p-1}/I^p depends only on their images in C and so I^{p-1}/I^p becomes naturally a sheaf of C-modules. It is clear that as a C-module $I^{p-1}/I^p|_{U_\alpha}$

$\cong C|_{U_\alpha} \otimes J^{p-1}/J^p$. Thus each I^{p-1}/I^p is locally free, and hence a vector bundle.

We now consider one case where the answer to our problem is immediate:

Proposition 2. Let D be a commutative algebra and $\varepsilon_0 : D \longrightarrow k$ a homomorphism with $J = \operatorname{Ker} \varepsilon_0$ satisfying $J^2 = 0$. If $\varepsilon : A \longrightarrow C$ is a D-manifold structure over X, then there is a class in $H^1(\operatorname{Der}(C,I))$ which vanishes if and only if there is a splitting $\varphi : C \longrightarrow A$ of ε. Here $I = I/I^2$ is given its natural C-module structure and Der(C,I) is the sheaf of derivations of C into I.

Proof. On $U_\alpha \cap U_\beta$ consider $\sigma_{\alpha\beta} : C \longrightarrow A$ defined by

$$\sigma_{\alpha\beta}(c) = \tau_\alpha^{-1}(c\otimes 1) - \tau_\beta^{-1}(c\otimes 1).$$

We have

$$\varepsilon(\sigma_{\alpha\beta}(c)) = c - c = 0,$$

so $\sigma_{\alpha\beta}$ has values in I, and

$$\begin{aligned}\sigma_{\alpha\beta}(c_1 c_2) &= \tau_\alpha^{-1}(c_1\otimes 1)\tau_\alpha^{-1}(c_2\otimes 1) - \tau_\beta^{-1}(c_1\otimes 1)\tau_\beta^{-1}(c_2\otimes 1)\\ &= \sigma_{\alpha\beta}(c_1)\tau_\alpha^{-1}(c_2\otimes 1) + \tau_\beta^{-1}(c_1\otimes 1)\sigma_{\alpha\beta}(c_2)\\ &= \sigma_{\alpha\beta}(c_1)\cdot c_2 + c_1\cdot\sigma_{\alpha\beta}(c_2),\end{aligned}$$

the last step using the definition of the C-module structure on I. Thus $\sigma_{\alpha\beta}$ is a 1-cocycle with values in Der(C,I). Its class in $H^1(\operatorname{Der}(C,I))$ is independent of the choice of the τ_α, as one can check.

Suppose this class vanishes, then, after possibly refining the cover $\{U_\alpha\}$, we have $\rho_\alpha \in \operatorname{Der}(C,I)(U_\alpha)$ with $\sigma_{\alpha\beta} = \rho_\alpha - \rho_\beta$. Thus we can define $\varphi : C \longrightarrow A$ by

$$\varphi(c) = \tau_\alpha^{-1}(c\otimes 1) - \rho_\alpha(c)$$

on U_α. Since the right hand sides agree on overlaps, φ is globally well-defined. φ is easily seen to be a homomorphism of algebras by reversing the calculation which showed $\sigma_{\alpha\beta}$ to be a derivation. Since $\varepsilon\rho_\alpha(c) = 0$, $\varepsilon\varphi(c) = \varepsilon\tau_\alpha^{-1}(c\otimes 1) = c$, so $\varepsilon\circ\varphi = \mathrm{id}_C$.

Conversely, given $\varphi : C \longrightarrow A$, then $\rho_\alpha(c) = \tau_\alpha^{-1}(c\otimes 1) - \varphi(c)$ defines a derivation of C into I with

$\sigma_{\alpha\beta} = \rho_\alpha - \rho_\beta$, showing that the class in $H^1(\mathrm{Der}(C,I))$ vanishes. □

The general case when $J^{n+1} = 0$ can be broken into a sequence of steps to which we can apply Proposition 2. To keep track of the local splittings through this sequence of steps we need the following algebraic Lemma:

Lemma 2. Let D_1, D_2, C be commutative k-algebras and

$$\varepsilon_i^\circ : D_i \longrightarrow k, \quad i = 1,2$$

homomorphisms of algebras with $J_i = \mathrm{Ker}\,\varepsilon_i^\circ$ nilpotent. Suppose $\pi_0 : D_1 \longrightarrow D_2$ is an epimorphism of algebras with $\varepsilon_2^\circ \circ \pi_0 = \varepsilon_1^\circ$. Put $A_i = C \otimes D_i$, $\varepsilon_i = \mathrm{id} \otimes \varepsilon_i^\circ$, $\pi = \mathrm{id} \otimes \pi_0$ and suppose $\varphi : C \longrightarrow A_2$ is a homomorphism of algebras satisfying $\varepsilon_2 \circ \varphi = \mathrm{id}$, then there is a homomorphism of algebras $\psi : C \longrightarrow A$ with $\pi \circ \psi = \varphi$, $\varepsilon_1 \circ \psi = \mathrm{id}$.

Proof. Define the algebra endomorphism $\Theta_2 : A_2 \longrightarrow A_2$ by $\Theta_2(c \otimes d) = \varphi(c) 1 \otimes d$. Then as in the proof of Proposition 1, Θ_2 is a unipotent endomorphism of A_2 and satisfies $\varepsilon_2 \circ \Theta_2 = \varepsilon_2$. It is also a morphism of A_2 as a D_2-module. By Lemma 1 $\delta_2 = \log \Theta_2$ is a nilpotent derivation of A_2 with $\varepsilon_2 \circ \delta_2 = 0$ and δ_2 is also a D_2-module morphism. Let $\sigma = \mathrm{id} \otimes \sigma_0$ where $\sigma_0 : D_2 \longrightarrow D_1$ is any linear map with $\pi_0 \circ \sigma_0 = \mathrm{id}$ and define $\delta_1 : A_1 \longrightarrow A_1$ by

$$\delta_1(c \otimes d) = \sigma(\delta_2(c \otimes 1)) 1 \otimes d$$

which is obviously a D_1-module morphism. Moreover

$$\delta_1(c_1 \otimes d_1 c_2 \otimes d_2) = \sigma(\delta_2(c_1 c_2 \otimes 1)) 1 \otimes d_1 d_2$$

$$= \sigma\{c_1 \otimes 1\, \delta_2(c_2 \otimes 1) + \delta_2(c_1 \otimes 1) c_2 \otimes 1\} 1 \otimes d_1 d_2$$

$$= c_1 \otimes d_1\, \delta_1(c_2 \otimes d_2) + \delta_1(c_1 \otimes d_1) c_2 \otimes d_2$$

since σ is C-linear. Thus δ_1 is a derivation of A_1 and

$$\pi(\delta_1(c \otimes d)) = \pi(\sigma(\delta_2(c \otimes 1)))\pi(1 \otimes d)$$

$$= \delta_2(c \otimes 1)\, 1 \otimes \pi_0(d)$$

$$= \delta_2(c \otimes \pi_0(d))$$

$$= \delta_2(\pi(c \otimes d)),$$

whilst

$$\varepsilon_1 \circ \sigma = \mathrm{id}\otimes(\varepsilon_1^\circ \circ \sigma_\circ) = \mathrm{id}\otimes(\varepsilon_2^\circ \circ \pi_\circ \circ \sigma_\circ) = \varepsilon_2$$

implies

$$\begin{aligned}\varepsilon_1(\delta_1(c\otimes d)) &= \varepsilon_1(\sigma(\delta_2(c\otimes 1)))\,\varepsilon_1^\circ(d)\\ &= \varepsilon_2(\delta_2(c\otimes 1))\,\varepsilon_1^\circ(d)\\ &= 0.\end{aligned}$$

Thus

$$\pi\circ\delta_1 = \delta_2\circ\pi, \quad \varepsilon_1\circ\delta_1 = 0.$$

From its definition δ_1 maps $C\otimes J_1^p$ into $C\otimes J_1^{p+1}$, so δ_1 is nilpotent and hence $\Theta_1 = \exp\delta_1$ is a unipotent endomorphism of D_1 with

$$\pi\circ\Theta_1 = \Theta_2\circ\pi, \quad \varepsilon_1\circ\Theta_1 = \varepsilon_1 .$$

If we set

$$\psi(c) = \Theta_1(c\otimes 1)$$

it is trivial to check that ψ satisfies the requirements of the Lemma. □

This Lemma gives us immediately the following:

<u>Proposition</u> 3. Let $\varepsilon : A \longrightarrow C$ be a D-manifold structure over X where D is commutative and suppose φ_p : $C \longrightarrow A/I^p$ splits $\varepsilon_p : A/I^p \longrightarrow C$. Then φ_p admits local liftings : $C|_{U_\alpha} \longrightarrow A/I^{p+1}|_{U_\alpha}$ splitting ε_{p+1}.

<u>Proof</u>. $\varphi_p : C \longrightarrow A/I^p$ transports via τ_α to give

$$\tilde\varphi_\alpha : C|_{U_\alpha} \longrightarrow C|_{U_\alpha}\otimes D/J^p.$$

The proof of Lemma 2 was an explicit algebraic construction, so applies to morphisms of sheaves of algebras to yield a lifting

$$\tilde\psi_\alpha : C|_{U_\alpha} \longrightarrow C|_{U_\alpha}\otimes D/J^{p+1}$$

of $\tilde\varphi_\alpha$ which transports back by τ_α to yield

$$\psi_\alpha : C|_{U_\alpha} \longrightarrow A/I^{p+1}|_{U_\alpha}$$

lifting φ_p . □

We are now in a position to attack the main problem. We no longer assume that D is commutative.

Proposition 4. Let $p \geq 1$. If $\varepsilon : A \longrightarrow C$ is a D-manifold structure over X and $\varphi_p : C \longrightarrow Z(A)/I_o^p$ a splitting, then there is a class in

$$H^1(\mathrm{Der}(C, I_o^p/I_o^{p+1}))$$

which vanishes if and only if $\varepsilon_{p+1} : Z(A)/I_o^{p+1} \longrightarrow C$ splits.

Proof. We can pull back

$$0 \longrightarrow I_o^p/I_o^{p+1} \longrightarrow Z(A)/I_o^{p+1} \xrightarrow{\pi_p} Z(A)/I_o^p \longrightarrow 0$$

by φ_p to give an extension

$$0 \longrightarrow I_o^p/I_o^{p+1} \longrightarrow E \longrightarrow C \longrightarrow 0 \qquad (1)$$

where

$$E = \{(a,c) \in Z(A)/I_o^{p+1} \oplus C : \pi_p(a) = \varphi_p(c)\}.$$

E is a subsheaf of algebras of $Z(A)/I_o^{p+1} \oplus C$ and I_o^p/I_o^{p+1} a sheaf of ideals of E with

$$(I_o^p/I_o^{p+1})^2 \subset I_o^{2p}/I_o^{p+1} = 0$$

since $p \geq 1$. Proposition 3 gives us local liftings of φ_p and these give us local splittings of (1). Note that the natural action of C on I_o^p/I_o^{p+1} given by (1) is the same as that given by $\varepsilon : A \longrightarrow C$. The local splittings of (1) then imply that $E \longrightarrow C$ is a

$$(k1 + J_o^p)/J_o^{p+1}$$

-manifold structure over X. Now apply Proposition 2. □

Corollary. If $H^1(\mathrm{Der}(C, I_o^p/I_o^{p+1})) = 0$ for all $p = 1,2,..$ then there is a homomorphism $\varphi : C \longrightarrow Z(A)$ splitting $\varepsilon : A \longrightarrow C$.

Proof. Since $Z(A)/I_o \cong C$, we have the identity map $\varphi_1 : C \longrightarrow Z(A)/I_o$ to start the induction going. Proposition 4 gives the inductive step to find $\varphi_p : C \longrightarrow Z(A)/I_o^p$ for each p. But $I_o^{n+1} = 0$ so $\varphi = {}^p\varphi_{n+1}$ will give the Corollary. □

Hence we have proven the following Theorem:

Theorem. Let D be a finite dimensional k-algebra, ε_0: $D \longrightarrow k$ a homomorphism of algebras such that $J = \text{Ker}\,\varepsilon_0$ is nilpotent. If $\varepsilon : A \longrightarrow C$ is a D-manifold structure over X such that $I_0 = \text{Ker}\,\varepsilon \cap Z(A)$ satisfies

$$H^1(\text{Der}(C, I_0^p/I_0^{p+1})) = 0$$

for all p, then there is a homomorphism $\varphi : C \longrightarrow Z(A)$ splitting ε, and as a sheaf of C-modules A is a vector bundle with structure group $\text{Aut}(D, \varepsilon_0)$.

Remark 1. $\text{Der}(C, I_0^p/I_0^{p+1})$ is a sheaf of C-modules which for $C = C^\infty$ is the tangent sheaf of X tensored over C with I_0^p/I_0^{p+1}. We do not in fact need to know this to conclude in many cases that the above cohomology groups vanish, for if C is a fine sheaf of algebras then for any sheaf of C-modules the higher cohomology groups all vanish by standard results in sheaf theory. In particular if $C = C^k$, $0 \leq k \leq \infty$, C is a fine sheaf if X is paracompact. Thus for any of these degrees of differentiability the conclusions of the theorem follow.

Remark 2. If $D = \Lambda \mathbb{R}^s$, then D has centre the even part of D. If $\text{Aut}_0(D, \varepsilon_0)$ denotes the subgroup of $\text{Aut}(D, \varepsilon_0)$ of $\mathbb{Z}_2$-degree preserving automorphisms, we easily see that if the τ_α are $\mathbb{Z}_2$-graded isomorphisms then a splitting of $\varepsilon : A \longrightarrow C$ into Z(A) gives an isomorphism of A as a sheaf of $\mathbb{Z}_2$-graded algebras with the sheaf of sections of a vector bundle of $\mathbb{Z}_2$ -graded algebras with structure group $\text{Aut}_0(D, \varepsilon_0)$. Remark 1 shows that such a splitting exists for $C = C^k$, $0 \leq k \leq \infty$. To prove Batchelor's theorem we have to show that the structure group of the vector bundle can be reduced to the subgroup of $\text{Aut}(D, \varepsilon_0)$ preserving the $\mathbb{Z}$-grading. This subgroup is canonically isomorphic to $GL(s, \mathbb{R})$, and passing from an element g of $\text{Aut}(D, \varepsilon_0)$ to its $\mathbb{Z}$-grading preserving part Gr(g) gives an epimorphism onto $GL(s, \mathbb{R})$ with a unipotent kernel. By Lemma 1 this kernel is diffeomorphic to its Lie algebra and so $\text{Aut}(D, \varepsilon_0)/GL(s, \mathbb{R})$ and $\text{Aut}_0(D, \varepsilon_0)/GL(s, \mathbb{R})$ are cells. We can thus always reduce from $\text{Aut}_0(D, \varepsilon_0)$ to $GL(s, \mathbb{R})$ in the C^k category.

Remark 3. Proposition 2 can be strengthened to show that if $J^2 = 0$, (isomorphism classes of) D-manifold structures $\varepsilon : A \longrightarrow C$ with a fixed vector bundle I as $\text{Ker}\,\varepsilon$ are in one-one correspondence with the classes in $H^1(\text{Der}(C, I))$. In the case X is a complex manifold and C the sheaf $\mathcal{O}$ of holomorphic functions then we see there are may be D-manifold structures over X which are not vector bundles. If $D = \Lambda \mathbb{R}^s$ then $I_0^p/I_0^{p+1} = \Lambda^{2p}E$, where $E = I/I^2$. Thus our obstructions are in $H^1(T \otimes \Lambda^{2p} E)$ with T the tangent sheaf. Green [3] also finds obstructions in the odd degrees; these are related to the fact that in the holomorphic category we cannot always reduce the structure group from $\mathrm{Aut}_o(D, \mathcal{E}_\bullet)$ to $GL(s,\mathbb{R})$.

References.

1. Batchelor,M. 1979, 'The structure of supermanifolds', Trans. Amer. Math. Soc. 253, pp.329-338.

2. Giachetti,R., Ragionieri,R. and Ricci,R. 1981, 'Symplectic structures on graded manifolds', J. Diff. Geom. 16, pp. 247-253.

3. Green,P. 1982, 'On holomorphic graded manifolds', Proc. Amer. Math. Soc. 85, pp. 587-590.

4. Kostant,B., 1977, 'Graded manifolds, Graded Lie Theory, and prequantization' in "Differential Geometric Methods in Mathematical Physics", Lecture Notes in Mathematics, 570, Springer-Verlag, Berlin, pp. 177-306.

5. Rothstein,M., Dissertation, UCLA.

CLASSICAL SUPERGRAVITY

James Isenberg*+

Department of Mathematics
University of Oregon
Eugene, OR 97403, U.S.A.

David Bao**‡

Institute for Advanced Studies
Princeton, NJ 08540, U.S.A.

Philip B. Yasskin

Department of Mathematics
Texas A & M University
College Station, TX 77843, U.S.A.

ABSTRACT

Most studies of supergravity concentrate on the quantum aspects of the theory. Here, we consider the properties of ($N = 1$) supergravity as a classical field theory; that is, as a system of partial differential equations on spacetime. We focus on the question of whether this system has a well-posed Cauchy problem. There is a consistent Hamiltonian (phase space) formulation of classical supergravity. However, proving well-posedness is not easy because the analytic techniques involved in such a proof require that one understand the function spaces in which the supergravity fields live - a task made difficult by the requirement that the spinor fields anticommute. We discuss an approach

---

* Research partially supported by NSF Grant MCS 83-03998

+ Current address: Dept. of Mathematics, Rice Univ., Houston, TX 77251

** Research partially supported by NSF Grant MCS 81-07086

‡ On leave from Dep. of Mathematics, Univ. of Houston, Houston, TX 7704

H.-J. Seifert et al. (eds.), Mathematical Aspects of Superspace, 173–205.

to this problem - that based on an infinite dimensional exterior algebra. The mathematics of this formulation is quite nice, and well-posedness is easily demonstrated. But the exterior algebra formulation results in a semi-decoupled, virtually linearized version of the field equations which does not seem very physical. There is as yet no known alternate formulation, but we do provide formal arguments which indicate that in any consistent formulation, classical supergravity has a well-posed Cauchy problem.

INTRODUCTION

The theory of supergravity [1], [2], [3], [4] has many striking features, including its invariance under a large supersymmetry group (which includes the spacetime diffeomorphisms as a subgroup), its mixing of boson fields and fermion fields (the supersymmetry group does this mixing), its consistent coupling of a high spin field (spin 3/2) to (a slightly modified version of) the Einstein gravitational field, and its reliance upon anti-commuting spinor fields. These features have stirred up considerable interest and hence a lot of work has gone into studying the possible role of supergravity and its various extensions in theoretical physics. Much is known, for example, about the particle representations these theories admit [3], [5], and about mechanisms for breaking their supersymmetry [6], [7]. There has also been extensive study of the possibility that supergravity is renormalizable [8].

In spite of all this interest in supergravity, there is one aspect of the theory which has received very little attention: its properties as a system of partial differential equations. (We shall use the term "classical supergravity" - abbreviated "CSG" - to refer to the study of such properties). Can solutions be constructed using Cauchy-Kowalesky methods? Is the Cauchy (i.e., initial value) problem well-posed? Do solutions contain causal cones restricting the propagation of disturbances? Are solutions stable in any sense? Are there in fact any nontrivial solutions?

In attempting to answer questions such as these using standard mathematical analysis, one must first be able to write the field equations as a system of partial differential equations on ordinary functions (living in an appropriate function space). For the Maxwell, Einstein, and Yang-Mills field theories, this is easily done - the components of those fields with respect to a chosen spacetime vector basis and a chosen gauge basis are ordinary functions, and so the field equations take the desired form. For supergravity, however, the components of the spinor field with respect to a chosen spacetime vector basis and spacetime spinor basis are <u>not</u> ordinary functions - they must anticommute. (The supergravity field equations are inconsistent unless this property holds). So unless something more is done, the supergravity field equations do not take the form of a system of partial differential equations on functions.

One thing one can do is choose an exterior ($\mathbb{Z}$-graded) algebra $\mathfrak{A}$ and define the supergravity fields so that they take values in $\mathfrak{A}$. This scheme builds into the fields the necessary anticommuting property and it casts the supergravity field equations into the desired (partial differential equation) form [see § 4]. Moreover, the resulting system of partial differential equations is well-behaved mathematically: it has a well-posed Cauchy problem, its solutions have a well-defined causal structure, and its solutions are stable.

There is one problem with this exterior algebra-valued formulation of CSG however: The fields decouple in a way which is incompatible with most notions of what the physical behavior of the classical gravitational field should be. This dilemma forces us to consider what, if any, physical relevance classical supergravity might have.

One of the main reasons why physicists have neglected the formal study of CSG is because most believe that if supergravity is to play a role in explaining physics, it will be as a quantum field theory. Hence their focus has been on quantum issues such as renormalizability rather than on classical issues like well-posedness. If in fact this belief is correct - so that (unlike Maxwell's and Einstein's theories) there is no direct classical field manifestation of supergravity - does it follow that CSG is irrelevant for physics?

We would like to argue no, because we believe (and experience with Yang-Mills theory shows) that classical solutions can be important tools for studying the behavior of quantum fields. They provide vacua and other critical points of the action about which one can do Feynman - type perturbations [9]. (Recall the Yang-Mills instantons). They serve as tools in carrying out the procedure of symmetry breaking (recall the Higgs mechanism [10]). They provide the basic arena for certain formulations of quantum field theory (for example, that of Segal [11]). Their characteristic surfaces govern the propagation of macroscopic signals (quantum as well as classical). And finally every quantum field theory must have a classical limit, and this most likely is related to the corresponding classical field theory (at least the bosonic part).

Of course there is the possibility that both quantum and classical supergravity have nothing at all to do with physics. Since that has not yet been shown, however, and since quantum field theory is not yet understood well enough to be able to dispense with assistance from the classical viewpoint, we believe that classical supergravity does warrant study.

This paper reports on the current status of research on CSG, with an emphasis on the dynamical analysis of the theory (i.e., on the initial value problem and well-posedness). We start with the formal definition of the supergravity field theory (N = 1 version, to keep the discussion simple) in § 1. In doing this, we take the spacetime point of view, so that supergravity is treated as a special coupling of a spin - 3/2 field to a gravitational tetrad (spin 2) field on spacetime rather than as a supertetrad field on superspacetime. This is the most convenient presentation for our purposes. We next discuss (in § 2) what the dynamical analysis is, using Maxwell theory as a simple example. In § 3, we do a formal dynamical analysis of supergravity. By "formal" we mean that the analysis does not depend upon an explicit function space definition of the supergravity fields (i.e., the spinor field components are simply treated as functions which anticommute). We next describe the exterior algebra formulation of CSG, in § 4. As mentioned above, this version of classical supergravity works very nicely mathematically, and we show this in § 4. We also point out the way in which the fields and equations decouple in the exterior algebra version, and how this leads to physically unpleasant behavior. This leads us back to the questions of physical interpretation of classical supergravity and to the possibility of other formulations, all of which are briefly discussed in § 5.

§ 1 DEFINITION OF CLASSICAL SUPERGRAVITY

Supergravity may be characterized as the only known consistent coupling of a massless spin-3/2 field and the Einstein gravitational field. This is especially noteworthy since, until 1976 (when supergravity was first formulated [1], [2]), it was thought that there was no way to consistently couple any field of spin-3/2 or greater to Einstein's theory. By a consistent coupling [12], we mean that the number of degrees of freedom of the nongravitational field (i.e., its "pure spin" character) is preserved, that any gauge freedom in the original theory survives the coupling, and that the integrability conditions for the field equations of the coupled theory are trivial. For some theories, such coupling is easily achieved: Klein-Gordon (spin 0), Dirac (spin 1/2), Maxwell (spin 1) and Yang-Mills (spin 1) are familiar examples. But for most others (Recall that one can formulate at least one classical field theory in flat spacetime for each integral and half-integral spin value) all attempts to attain a consistent coupling have failed.

To make supergravity work, two rather surprising conditions must be satisfied by the fields: (1) The connection field must contain torsion (which is determined as an algebraic function of the other fields via the field equations), and (2) the components of the spinor field must anticommute. Without these two conditions (the Majorana condition on the spinor field is needed as well) consistency fails. Hence it is not surprising that it took so long to find a consistent coupling of spin-3/2 to gravity.

We now formally set up the N = 1 supergravity field theory. To simplify calculations in later sections, we shall eschew the fancier formulations* and primarily use "index cruncher" language (with occasional expressions being written in differential form language). Our conventions on indices, on γ matrices, and on geometric objections are summarized in the Appendix. All fields are defined on a spacetime manifold M

<u>Tetrad Field</u> (spin 2)

$$e^{\hat{\alpha}} = e^{\hat{\alpha}}{}_{\mu}\, dx^{\mu} \,. \tag{1.1}$$

This is an $\mathbb{R}^4$-valued one-form field (i.e., a collection of four linearly independent one-form fields). It specifies the spacetime metric via

$$g_{\mu\nu} = \eta_{\hat{\alpha}\hat{\beta}}\, e^{\hat{\alpha}}{}_{\mu} \otimes e^{\hat{\beta}}{}_{\gamma} \tag{1.2}$$

where $\eta_{\hat{\alpha}\hat{\beta}}$ is the Minkowski metric. The dual to the tetrad field is denoted by $e_{\hat{\beta}}{}^{\mu}$, and it satisfies

$$e_{\hat{\alpha}}{}^{\mu}\, e^{\hat{\alpha}}{}_{\nu} = \delta^{\gamma}{}_{\mu} \tag{1.3a}$$

as well as

$$e_{\hat{\alpha}}{}^{\mu}\, e^{\hat{\beta}}{}_{\mu} = \delta_{\hat{\alpha}}{}^{\hat{\beta}} \tag{1.3b}$$

Note that "tetrad indices" (like $\hat{\alpha}$) are raised and lowered using $\eta_{\hat{\alpha}\hat{\beta}}$ while "tangent indices" (like μ) are raised and lowered using $g_{\mu\nu}$. Also, on covariant objects (<u>not</u> including the connection field) one may freely swap between tetrad and tangent indices using $e^{\hat{\alpha}}{}_{\mu}$ and $e_{\hat{\beta}}{}^{\nu}$.

<u>Spinor Field</u> (Spin 3/2)

$$\psi^{A} = \psi^{A}{}_{\mu}\, dx^{\mu}. \tag{1.4}$$

This is a Majorana spinor-valued one form which anticommutes. The Majorana condition, in general, requires that the charge conjugate spinor $\hat{\psi}^{A}{}_{\mu} := \psi^{B}{}_{\mu}\, C_{B}{}^{A}$ (where $C_{B}{}^{A}$ is the charge conjugation matrix) be equal to $\psi^{A}{}_{\mu}$ itself. In the real representation of the γ-algebra (see Appendix) it implies that the components of $\psi^{A}{}_{\mu}$ are all real-valued (rather than complex valued). The anticommutation condition requires that the components of a pair of the spinor fields behave

---

* See [13] for a discussion of the superspace formulation. See [14] for a principal bundle formulation of the Einstein-Dirac theory. This serves as a prototype for such a formulation of supergravity.

as follows*:

$$\psi^A{}_\mu \, \psi^B{}_\nu = - \psi^B{}_\nu \, \psi^A{}_\mu . \tag{1.5}$$

Note that we shall drop the spinor index "A" on $\psi^A{}_\mu$ whenever this leads to no confusion.

Connection Field

$$\Gamma^{\hat\alpha\hat\beta} = \Gamma^{\hat\alpha\hat\beta}{}_\mu \, dx^\mu \tag{1.6}$$

This is a Lorentz algebra-valued one form field which specifies the spacetime connection. The connection is metric-compatible (as guaranteed by the condition $\Gamma^{\hat\alpha\hat\beta}{}_\mu = -\Gamma^{\hat\beta\hat\alpha}{}_\mu$) but generally not torsion free:

$$Q^{\hat\alpha}{}_{\mu\nu} := \partial_{[\mu} \, e^{\hat\alpha}{}_{\nu]} + \Gamma^{\hat\alpha}{}_{\hat\beta[\mu} \, e^{\hat\beta}{}_{\nu]} \neq 0 . \tag{1.7}$$

One can write

$$\Gamma^{\hat\alpha\hat\beta}{}_\mu = \{^{\hat\alpha\hat\beta}{}_\mu\} - Q^{\hat\beta\hat\alpha}{}_\mu + Q_\mu{}^{\hat\beta\hat\alpha} - Q^{\hat\alpha}{}_\mu{}^{\hat\beta} \tag{1.8}$$

where $\{^{\hat\alpha\hat\beta}{}_\mu\}$ denotes the Riemannian connection: metric compatible and torsion-free, a simple function of the tetrad and its derivatives.

The connection $\Gamma^{\hat\alpha\hat\beta}{}_\mu$ defines a covariant derivative ∇_μ on the various fields on spacetime. To make it explicit, we need the tangent index form of the connection**

$$\Gamma^\xi{}_{\mu\nu} = \Gamma^{\hat\alpha}{}_{\hat\beta\nu} \, e^{\hat\beta}{}_\mu \, e_{\hat\alpha}{}^\xi + \partial_\nu (e^{\hat\alpha}{}_\mu) e_{\hat\alpha}{}^\xi . \tag{1.9}$$

Then, for the spinor field $\psi^A{}_\mu$, we have

$$\nabla_\nu \psi^A{}_\mu = \partial_\nu (\psi^A{}_\mu) + \frac{1}{2} \Gamma^{\hat\alpha\hat\beta}{}_\nu \, \sigma_{\hat\alpha\hat\beta}{}^A{}_B \psi^B{}_\mu - \Gamma^\lambda{}_{\mu\nu} \, \psi^A{}_\lambda \tag{1.10}$$

---

* Do not confuse the anti-commutation of the components of ψ^A with that of the wedge product of two one-form fields. Indeed, we find that for $\psi^A := \psi^A{}_\mu \, dx^\mu$ we have $\psi^A \wedge \psi^B = \psi^B \wedge \psi^A$ because two minus signs are involved.

** This is just the usual basis transformation form for connection coefficients, as applied to the transformation from the tetrad basis e^α to the tangent basis dx^μ.

while for an object such as the torsion tensor $Q^{\hat{\alpha}}{}_{\mu\nu}$, we have

$$\nabla_\lambda Q^{\hat{\alpha}}{}_{\mu\nu} = \partial_\lambda Q^{\hat{\alpha}}{}_{\mu\nu} + \Gamma^{\hat{\alpha}}{}_{\hat{\beta}\lambda} Q^{\hat{\beta}}{}_{\mu\nu} - \Gamma^{\sigma}{}_{\mu\nu} Q^{\hat{\alpha}}{}_{\sigma\nu} - \Gamma^{\sigma}{}_{\nu\lambda} Q^{\hat{\alpha}}{}_{\mu\sigma}. \quad (1.11)$$

For the tetrad field, one finds

$$\nabla_\nu e^{\hat{\alpha}}{}_\mu = \partial_\nu e^{\hat{\alpha}}{}_\mu + \Gamma^{\hat{\alpha}}{}_{\hat{\beta}\nu} e^{\hat{\beta}}{}_\mu - \Gamma^{\xi}{}_{\mu\nu} e^{\hat{\alpha}}{}_\xi = 0 \quad (1.12)$$

which re-expresses equation (1.9).

For many purposes (mostly related to the one-form nature of the various fields) it is useful to define a slightly modified version of ∇_μ, which we shall label as D_ν. This operator acts properly on tetrad and spinor indices, but it ignores tangent indices. For example

$$D_\nu \psi^A{}_\mu = \partial_\nu \psi^A{}_\mu + \frac{1}{2} \Gamma^{\hat{\alpha}\hat{\beta}}{}_\nu \sigma_{\hat{\alpha}\hat{\beta}}{}^A{}_B \psi^B{}_\mu \quad (1.13)$$

and

$$D_\nu e^{\hat{\alpha}}{}_\mu = \partial_\nu e^{\hat{\alpha}}{}_\mu + \Gamma^{\hat{\alpha}}{}_{\hat{\beta}\nu} e^{\hat{\beta}}{}_\mu = \Gamma^{\xi}{}_{\mu\nu} e^{\hat{\alpha}}{}_\xi \neq 0. \quad (1.14)$$

This operator is not strictly covariant, but the closely related object "D" is. D is the exterior covariant differential operator, the action of which on the one-form ψ^A is given by

$$D\psi^A = d\psi^A + \frac{1}{2} \Gamma^{\hat{\alpha}\hat{\beta}} \sigma_{\hat{\alpha}\hat{\beta}}{}^A{}_B \psi^B \quad (1.15)$$

where d is the usual exterior differential, whose index form is

$$(d\psi^A)_{\mu\nu} = \partial_\mu \psi^A{}_\nu - \partial_\nu \psi^A{}_\mu . \quad (1.16)$$

Note that

$$De^{\hat{\alpha}} = de^{\hat{\alpha}}{}_\mu + \Gamma^{\hat{\alpha}}{}_{\hat{\beta}} \wedge e^{\hat{\beta}} = 2Q^{\hat{\alpha}} \quad (1.17)$$

which is the torsion two-form.

Of course the connection, besides determining spacetime torsion and the covariant derivative on spacetime fields, determines spacetime curvature. In form language, the curvature is given by the two-form

$$\Omega^{\hat{\alpha}}{}_{\hat{\beta}} = d\Gamma^{\hat{\alpha}}{}_{\hat{\beta}} + \Gamma^{\hat{\alpha}}{}_{\hat{\lambda}} \wedge \Gamma^{\hat{\lambda}}{}_{\hat{\beta}} \quad (1.18)$$

the components of which are the familiar Riemann tensor

$$\Omega^{\hat{\alpha}}{}_{\hat{\beta}} = \frac{1}{2} R^{\hat{\alpha}}{}_{\hat{\beta}\mu\nu} dx^\mu \wedge dx^\nu . \quad (1.19)$$

From the Riemann tensor, one defines the Ricci tensor by

$$R^{\hat{\alpha}}{}_{\hat{\beta}} := R^{\alpha}{}_{\beta\mu\nu}\, e_{\hat{\alpha}}{}^{\mu} \tag{1.20}$$

(with $R_{\mu\nu} = e^{\hat{\beta}}{}_{\mu}\, R_{\hat{\beta}\nu}$ generally <u>not</u> symmetric), the scalar curvature by

$$R = R_{\hat{\beta}\nu}\, e^{\hat{\beta}\nu} \tag{1.21}$$

and the Einstein tensor by

$$G^{\hat{\beta}}{}_{\mu} = R^{\hat{\beta}}{}_{\mu} - \frac{1}{2}\, e^{\hat{\beta}}{}_{\mu}\, R\ . \tag{1.22}$$

The usual relation between the curvature and the commutation of covariant derivatives holds: on $\psi^{A}{}_{\mu}$ it takes the form

$$\nabla_{[\nu}\,\nabla_{\lambda]}\,\psi^{A}{}_{\mu} = \frac{1}{2}\, R^{\hat{\alpha}\hat{\beta}}{}_{\nu\lambda}\,\sigma_{\hat{\alpha}\hat{\beta}}{}^{A}{}_{B}\,\psi^{B} - R^{\xi}{}_{\mu\nu\lambda}\,\psi^{A}{}_{\xi}\ . \tag{1.23}$$

Using the D operator, we get

$$D^2\ \psi^{A} = \frac{1}{2}\,\Omega^{\hat{\alpha}\hat{\beta}}\,\sigma_{\hat{\alpha}\hat{\beta}}{}^{A}{}_{B}\,\psi^{B}\ . \tag{1.24}$$

The action for supergravity is obtained by minimally coupling the Rarita-Schwinger spin-3/2 action* [15] to the Einstein-Cartan action. (This is just the Einstein action, but with torsion in the connection). It is**

$$S_{SG}\,[e^{\hat{\alpha}},\psi,\Gamma^{\hat{\alpha}\hat{\beta}}] = \int_{M} \left[\frac{1}{2}\, R\ e + \frac{1}{2}\,\bar{\psi}\ \gamma_5\wedge\gamma\wedge D\psi\right]$$

$$= \int_{M} d^4x\ \sqrt{-g}\left[\frac{1}{2}R + \frac{1}{2}\varepsilon^{\lambda\mu\nu\rho}\ \bar{\psi}_{\lambda}\ \gamma_5\gamma_{\mu}D_{\nu}\psi_{\rho}\right] \tag{1.25}$$

where e is the volume element 4-form constructed from the tetrad $e^{\hat{\alpha}}$, $\gamma_5 := \gamma^{\hat{0}}\,\gamma^{\hat{1}}\,\gamma^{\hat{2}}\,\gamma^{\hat{3}}$, $\gamma := \gamma_{\hat{\alpha}}\, e^{\hat{\alpha}}$ (a matrix-valued one-form), $\sqrt{-g} := [-\det(g_{\mu\nu})]^{1/2}$ and $\varepsilon^{\lambda\mu\nu\rho}$ is the totally skew Levi-Civita tensor (not tensor density).

---

* Note that the Rarita-Schwinger theory in flat spacetime can be formulated with or without the Majorana condition and with or without the fields anticommuting.

** We leave out the boundary terms. This may be justified by assuming that $M = \Sigma \times \mathbb{R}$, where Σ is compact without boundary.

To obtain the supergravity field equations, one varies the action with respect to the three fields $e^{\hat\alpha}{}_\mu$, $\psi^A{}_\mu$, and $\Gamma^{\alpha\beta}$. The equations are

$$G^\mu{}_{\hat\alpha} = \frac{1}{2}\varepsilon^{\mu\lambda\nu\rho}\,\bar\psi_\lambda\,\gamma_5\,\gamma_{\hat\alpha}\,D_\nu\,\psi_\rho\ , \tag{1.26}$$

$$\varepsilon^{\lambda\mu\nu\rho}\,\gamma_\mu\,D_\nu\,\psi_\rho = 0\ , \tag{1.27}$$

and $$Q^\xi{}_{\mu\nu} = +\,\frac{1}{4}\,\bar\psi_\mu\,\gamma^\xi\,\psi_\nu\ . \tag{1.28}$$

The first of these is the familiar Einstein equation with stress-energy tensor $T^\mu{}_{\hat\alpha} = \varepsilon^{\mu\lambda\nu\rho}\,\bar\psi_\lambda\,\gamma_5\,\gamma_{\hat\alpha}\,D_\nu\,\psi_\rho$ (and with torsion). The second is the curved spacetime version of the Rarita-Schwinger equation for the spin-3/2 field. For later purposes, it is useful to have an alternate version (1.27). If we define the operator i_γ on any form A as

$$i_\gamma\,A := \gamma^{\hat\alpha}\,ie_{\hat\alpha}\,A = \gamma^{\hat\alpha}\,A_{\hat\alpha} \tag{1.29}$$

(so $i_\gamma A$ is "insertion of the vector $\gamma^{\hat\alpha}\,e_{\hat\alpha}$ into the first slot of the form A), then one can show [16] that

$$\begin{aligned} 0 &= i_\gamma\,D\psi \\ &= \gamma^\mu(D_\mu\psi_\lambda - D_\lambda\psi_\mu) \end{aligned} \tag{1.30}$$

is exactly equivalent to (1.27).

The last of the field equations, (1.28), determines the torsion as a function of the other fields (This sort of behavior always occurs in field theories which, like supergravity, have an Einstein-Cartan type of action). We may use equation (1.28) to replace $\Gamma^{\hat\alpha\hat\beta}{}_\mu$ everywhere throughout the other equations by

$$\Gamma^{\hat\alpha\hat\beta}{}_\mu = \{{}^{\hat\alpha\hat\beta}{}_\mu\} + \frac{1}{4}(-\bar\psi_\mu\gamma^{\hat\alpha}\psi^{\hat\beta} + \bar\psi^{\hat\beta}\gamma_\mu\psi^{\hat\alpha} - \bar\psi^{\hat\alpha}\gamma^{\hat\beta}\psi_\mu) \tag{1.31}$$

Hence, supergravity essentially involves only the equations (1.26) to (1.30) and only the fields $e^{\hat\alpha}{}_\mu$ and $\psi^A{}_\mu$.

We noted earlier that the supergravity field equations are consistent. One manifestation of this is that the integrability conditions for the system (1.26) and (1.30) are automatically satisfied. That is, demanding that the derivatives of equations (1.26) and (1.30) vanish introduces no new conditions of the fields. While we will not go through the analysis (see, e.g. [17]), we do wish to remind the reader that without the Majorana condition, the anticommutation condition, and the effects of torsion, this consistency fails.

In verifying consistency, one finds that certain conservation laws play an important role. These conservation laws are closely tied, via Noether's theorem, to the invariance of the supergravity action S (and the consequent covariance of the supergravity field equations) under certain transformations of the fields. One has the usual spacetime diffeomorphism group and Lorentz frame rotation group of general relativity. In addition, one finds that $S_{SG}[e,\psi,\Gamma]$ is invariant under the infinitesmal supersymmetry transformations

$$\delta e^{\hat{\alpha}} = \frac{1}{2}\bar{\xi}\,\gamma^{\hat{\alpha}}\,\psi \tag{1.32a}$$

$$\delta\psi = D\xi \tag{1.32b}$$

and $$\delta\Gamma^{\hat{\alpha}\hat{\beta}} = \frac{1}{2}\,\xi\,\gamma_5[\gamma*(e^{\hat{\alpha}}\wedge e^{\hat{\beta}}\wedge D\psi) + e*(e^{\hat{\beta}}\wedge\gamma\wedge D\psi)] \tag{1.32c}$$

where ξ is an (infinitesmal) anti-commuting Majorana spinor field (not a one-form) which generates the transformation ("*" is the Hodge dual map on forms). These transformations, which map solutions to solutions, mix the bosons (e^{α}) and fermions (ψ^A) of the theory. This is a key feature of supergravity. Note that there is as yet no known finite representation of the supersymmetry transformation of the fields.

In the discussions of this section, we have ignored the problem of finding a precise mathematical definition of the fields which permits them to have the properties indicated (including anticommutation of the components $\psi^A{}_\mu$ of the spin-3/2 field). We shall continue to ignore this problem, working formally, until § 4.

§ 2 DYNAMICAL ANALYSIS OF CLASSICAL FIELD THEORIES.

One way to obtain solutions of a given set of field equations is via the Cauchy problem: One chooses the fields at some initial time t = 0 and then uses the field equations to evolve them throughout spacetime.

For classical field theories with lots of gauge freedom (like supergravity) one must do a bit of work before the theory is in a form which is amenable to the Cauchy method. This work, which we call the dynamical analysis of the theory, involves the following steps: (1) One replaces the spacetime covariant fields Z by fields z which are spatially covariant (relative to a spacelike foliation of spacetime). (2) One constructs a phase space for the theory involving the spatial fields z and their conjugate momenta p. (3) One recognizes some of the field equations to be constraints $\Phi(z,p) = 0$ on the possible choice of initial data $(z,p)|_{t=0}$. (4) One casts the rest of the field equations into the form of evolution equations which specify the time derivatives of the fields as functionals of the fields:

$$\partial_t z = F(z,p) \qquad \partial_t p = G(z,p) \ . \tag{2.1}$$

(5) One checks that the constraint equations are preserved by the evolution equations, in the sense that

$$0 = \partial_t \Phi = D_z\Phi\ \partial_t z + D_p\Phi\ \partial_t p = D_z\Phi \cdot F + D_p\Phi \cdot G \ . \tag{2.2}$$

Often, to guarantee that the constraints of a given theory are preserved, one may have to add more constraints to the original set. In some cases, however, no number of additional constraints is enough to guarantee preservation. In such a case (Rarita-Schwinger coupled to Einstein gravity <u>without torsion</u> is an example) the theory is inconsistent.

Even if the theory is consistent, and all of the above steps can be successfully carried out, one of the most important steps remains. This is the proof that the initial value formulation is well-posed. Specifically, one must prove mathematically that any choice of initial data which satisfies the constraints can be evolved (via the evolution equations) into a nonsingular solution for some finite amount of time into the future (or past). If well-posedness cannot be proven, then the Cauchy formulation of the theory is worth little.

Before sketching out a brief example of the dynamical analysis (for the Maxwell theory) we note an interesting division of labor among the fields which it uncovers. For some of the fields, the time development is <u>not</u> determined by the evolution equations. These, which never appear in the constraints either, we call the "atlas fields". The rest of the fields, the time development of which <u>is</u> controlled by the evolution equations, we label the "dynamic fields". The atlas fields, in effect, control the gauge transformations (or atlas) during evolution [18]. One may choose them freely. The dynamic fields, in effect, carry the physical field information (along with some gauge information mixed in). They are determined completely by the initial data (and the choice of atlas fields throughout time).

Let us now consider the Maxwell field theory in flat spacetime. The spacetime covariant formulation is based upon the electromagnetic potential A, which is a spacetime one-form $A = A_\mu\, dx^\mu$. From A, one constructs the electromagnetic field $F = dA$, which in component form is $F_{\mu\nu} = \partial_\mu A_\nu - \partial_\nu A_\mu$. The action is

$$\underset{M}{S}[A] = -\frac{1}{4}\int_M F\wedge *F = -\frac{1}{4}\int_M d^4x\ F_{\mu\nu}F^{\mu\nu} \tag{2.3}$$

and the field equation is

$$0 = \partial_\mu F^\mu{}_\nu \ . \tag{2.4}$$

For the dynamic analysis, one replaces A by $a = a_i\, dx^i$, a space-covariant one-form, and $\theta := i_{\partial_t} A = A_0$, a spatial scalar. One may rewrite the action in terms of these fields and their derivatives

$$S_M[a,\theta] = \int dt \int_\Sigma d^3x\, L_M = \int dt \int_\Sigma d^3x \left[\frac{1}{2}(\dot a_i - \partial_i\phi)(\dot a^i - \partial^i\phi) - \frac{1}{2}\, {}^3F_{ij}\, {}^3F^{ij}\right], \qquad (2.5)$$

(where "$\cdot$" refers to ∂_t) and then do the rest of the analysis via the Bergman-Dirac method [19], [20]. That is, one obtains the momenta as

$$E^i := \frac{\delta L}{\delta \dot a_i} = \dot a^i - \partial^i\theta \qquad p := \frac{\delta L}{\delta\dot\theta} = 0 \,. \qquad (2.6)$$

(Note the primary constraint $p = 0$), one obtains the Hamiltonian as

$$H = \int_\Sigma d^3x\, (E^i\, \dot a_i + p\dot\theta - L + \mu p)$$
$$= \int_\Sigma d^3x\, \left(\frac{1}{2} E^i\, E_i + \frac{1}{2}\, {}^3F^{ij}\, {}^3F_{ij} - \theta(\partial_i\, E^i) + \mu p\right) \qquad (2.7)$$

(with μ as a Lagrange multiplier) and the evolution equations are then

$$\partial_t\, a_i = \frac{\delta H}{\delta E^i} = E_i + \partial_i\theta \qquad \partial_t\theta = \frac{\delta H}{\delta p} = \mu \qquad (2.8)$$
$$\partial_t\, E^i = \frac{-\delta H}{\delta a_i} = -\,\partial_i\, {}^3F^{ij} \qquad \partial_t\, p = \frac{-\delta H}{\delta\theta} = +\,\partial_i\, E^i \,.$$

Now if one demands that $\partial_t\, p = 0$, one obtains a new constraint

$$0 = \partial_i\, E^i \,. \qquad (2.9)$$

Preserving this, one finds

$$0 = \partial_t(\partial_i\, E^i) = \partial_i\partial_t\, E^i = -\,\partial_i\partial_j\, {}^3F^{ij} = 0 \qquad (2.10)$$

so the theory is consistent.

Note that the Lagrange multiplier μ is completely free, and so the evolution of θ is not fixed by (2.8). Hence θ is an atlas field: one may choose it freely, and by choosing it one fixes the time development of the gauge choice.

To prove that the Cauchy problem for the Maxwell theory is

well-posed, it is useful to go back to the spacetime covariant formulation of the theory. A key result [21] from the theory of hyperbolic partial differential equations states, roughly, that for any second order system of pde's of the hyperbolic form

$$L^{\mu\nu}[Z]\partial_\mu \partial_\nu Z = 0_1 \qquad (2.11)$$

there are unique local (in time) solutions to the corresponding Cauchy problem, as long as the Cauchy data lives in certain function spaces (We shall refer to this as the Leray theorem). Here $L^{\mu\nu}[Z]$ must be a Lorentz metric, possibly depending upon Z and its first derivatives, and 0_1 is a function which also may involve Z and its first derivatives. It is rather remarkable that not only local existence but causal propagation as well is guaranteed by this theorem (with $L^{\mu\nu}$ determining the causal cone structure).

The Maxwell field equation

$$0 = \partial_\mu F^\mu{}_\nu = \partial_\mu \partial^\mu A_\nu - \partial_\nu(\partial^\mu A_\mu) \qquad (2.12)$$

does not take the hyperbolic form (2.11). However, we can argue as follows that the Leray theorem still guarantees that its Cauchy problem is well-posed: Consider the second order system

$$0 = \partial_\mu \partial^\mu A_\nu \,. \qquad (2.13)$$

This is of the proper form, so its Cauchy problem is well-posed. Now note that if A_μ solves (2.13) and if it also satisfies the condition

$$\partial^\mu A_\mu = 0 \qquad (2.14)$$

then it solves the Maxwell equation as well. Thus one way to obtain a solution to the Maxwell equation (2.11) is to choose initial data for Maxwell, evolve via (2.12), and hope that $\partial^\mu A_\mu = 0$ is satisfied. But for <u>every</u> choice of Maxwell initial data, this can be made to work (via choice of gauge). This is because, first of all, a choice of Maxwell initial data $\{a_i, E^j\}_{|t=0}$ does not fix $\partial_t A_o$ (As seen above, $A_o = \theta$ is an atlas field the time evolution of which is fixed neither by the choice of Maxwell initial data nor by the Maxwell evolution equations). Therefore, for any choice $\{a_i, E^j\}_{|t=0}$, one can choose $\partial_t A_{o|t=0}$ so that the extension to initial data $\{A_\mu, \partial_t A_\nu\}_{|t=0}$ for the system (2.13) satisfies $\partial^\mu A_{\mu|t=0}$ and $\partial_t(\partial^\mu A_\mu)_{|t=0} = 0$*.

* The choice to make is $\partial_t A_{o|t=0} = \partial^i A_{i|t=0} = \partial^i a_i$. Then $\partial^\mu A_{\mu|t=0} = 0$ follows immediately. As for $\partial_t(\partial^\mu A_\mu)_{|t=0}$, one uses (2.13) and the constraint $\partial_i E^i = 0$ to show that it vanishes:
$\partial_t(\partial^\mu A_\mu) = -\partial_t^2 A_o + \partial^i \partial_t A_i = -\partial^i \partial_i A_o + \partial^i(E_i + \partial_i A_o) = 0$.

Now, it is a consequence of (2.13) that the quantity $\partial^{\nu}A_{\nu}$ satisfies

$$\partial_{\mu}\,\partial^{\mu}(\partial^{\nu}A_{\nu}) = 0 \tag{2.15}$$

(just take the divergence of equation (2.13)), which is a hyperbolic equation of the form (2.11). Hence, the existence and uniqueness of solutions to (2.15), together with the initial data (all vanishing) require that as A_{μ} is evolved via (2.13), it always satisfies $\partial^{\mu}A_{\mu} = 0$. This gives us the desired existence of solutions for the Maxwell Cauchy problem, and it gives us causal propagation for the Maxwell theory as well.

Note that one can prove that <u>every</u> solution of the Maxwell equations can be obtained via the Cauchy problem. To do this, one need only show that, for any given solution A_{μ} ther exists a gauge transformation which maps A_{μ} to $\widetilde{A}_{\mu}$ such that $\partial^{\mu}\widetilde{A}_{\mu} = 0$. For Yang-Mills and for Einstein's theory*, one can show that as well. We will see that for supergravity, however, no such proof yet exists. (The big problem being the lack of well-defined finite supersymmetry transformations).

There are of course other ways to study the field equations of a classical field theory besides that just sketched. However, the dynamical analysis has the following important features: (1) It provides a straightforward way to determine whether or not a given theory is consistent. (2) It reveals the gauge freedom in the solutions and provides one with a way to control it (the atlas fields). (3) It determines the number of degrees of freedom in the theory. (4) It determines whether or not the theory has a causal cone structure restricting the propagation of signals. (5) It provides one with a procedure for generating solutions. (6) It serves as a starting point for the Fischer-Marsden-Moncrief [22] analysis of the linearization stability of solutions and the manifold structure of the space of all solutions. (7) It formulates the theory in canonical form, which is one of the jumping off points for quantization.

§ 3 FORMAL DYNAMICAL ANALYSIS OF CLASSICAL SUPERGRAVITY.

In this section, we discuss the dynamical formulation of classical supergravity, and give a formal argument that the Cauchy problem for CSG is well-posed. Since the details of the dynamical formulation are quite messy, and since they appear elsewhere [23], [24], and since one can understand much of what is going on without them, we shall leave many of the details out. Most of our attention will be focussed on the proof of well-posedness.

---

* Globally hyperbolic solutions only.

We start by decomposing the fields. Recall that all of the supergravity fields $e^{\hat{\alpha}}$, ψ^A, and $\Gamma^{\hat{\alpha}\hat{\beta}}$ are "___"-valued one-forms. We do not decompose the "___"; all we do is replace each "___"-valued spacetime one-form by a "___"-valued spatial one-form and a "___"-valued spatial scalar. Specifically, if Σ is an embedded spacelike surface $I: \Sigma \to M$, and if ∂_t is a transverse timelike vector field, then we have*

$$e^{\hat{\alpha}} \to \begin{pmatrix} E^{\hat{\alpha}} := I^* e^{\hat{\alpha}} = e^{\hat{\alpha}}{}_i \, dx^i \\ N^{\hat{\alpha}} := i_{\partial_t} e^{\hat{\alpha}} = e^{\hat{\alpha}}{}_0 \end{pmatrix}, \tag{3.1}$$

$$\psi^A \to \begin{pmatrix} \phi^A := I^* \psi^A = \psi^A{}_i \, dx^i \\ \lambda^A := i_{\partial_t} \psi^A = \psi_0{}^A \end{pmatrix}, \tag{3.2}$$

and
$$\Gamma^{\hat{\alpha}\hat{\beta}} \to \begin{pmatrix} \Lambda^{\hat{\alpha}\hat{\beta}} := I^* \Gamma^{\hat{\alpha}\hat{\beta}} = \Gamma^{\hat{\alpha}\hat{\beta}}{}_i \, dx^i \\ \ell^{\hat{\alpha}\hat{\beta}} := i_{\partial_t} \Gamma^{\hat{\alpha}\hat{\beta}} = \Gamma^{\hat{\alpha}\hat{\beta}}{}_0 \end{pmatrix}. \tag{3.3}$$

It is interesting to note, before going through the analysis, that $N^{\hat{\alpha}} = e^{\hat{\alpha}}{}_0$, $\lambda^A = \psi^A{}_0$, and $\ell^{\hat{\alpha}\hat{\beta}} = \Gamma^{\hat{\alpha}\hat{\beta}}{}_0$ turn out to be the atlas fields of supergravity, just as $\theta = A_o$ is the atlas field of Maxwell's theory.

Using the spatial fields defined above, one can rewrite the action S_{SG} (see equation (1.25)) in the 3 + 1 form

$$S_{SG}[E^{\hat{\alpha}}, N^{\hat{\alpha}}, \phi^A, \lambda^A, \Lambda^{\hat{\alpha}\hat{\beta}}, \ell^{\hat{\alpha}\hat{\beta}}] = \int dt \int_\Sigma d^3x \; L_{SG} \tag{3.4}$$

analogous to (2.5) for Maxwell's theory. This is the first of many expressions the detailed form of which we shall leave out, due to its complexity. Then from (3.4), one calculates the conjugate momenta for the six fields. The momentum conjugate to $N^{\hat{\alpha}}, \lambda^A, \ell^{\hat{\alpha}\hat{\beta}}$, and $\Lambda^{\hat{\alpha}\hat{\beta}}$ (we call them $p_{(N)\hat{\alpha}}$, $p_{(\lambda)A}$, $p_{(\ell)\hat{\alpha}\hat{\beta}}$, and $p_{\Lambda\hat{\alpha}\hat{\beta}}$, respectively) all vanish** and thus introduce constraints. Those conjugate to $E^{\hat{\alpha}}$ and ϕ^A do not, although for the spinor field we get

* Note that the pull-back map I^* is essentially a spatial projection while i_{∂_t} is a timelike projection. We have been less than careful regarding whether these fields live on Σ or M, because that is not important for the discussion here. See [14] for a more careful treatment.

** Actually, p_Λ vanishes only if one performs an integration-by-parts on the scalar curvature term in the action before doing any variation.

$$\underset{(\phi)}{p}{}^{i}_{A} = \frac{\delta L_{SG}}{\delta \dot{\phi}^{A}_{i}} = \sqrt{{}^3g}\ \bar{\phi}_j\ \gamma_{\perp}\ \sigma^{ji} \tag{3.5}$$

which is another constraint (here ${}^3g$ is the determinant of the spatial metric ${}^3g_{ij} = \eta_{\hat{\alpha}\hat{\beta}}E^{\hat{\alpha}}{}_i\ E^{\hat{\beta}}{}_j$, and $\gamma_{\perp} := \gamma_{\mu}\ n^{\mu}$ where n^{μ} is the vector field* normal to the surface Σ.

The next few steps are to write out the Hamiltonian, to find the evolution equations (by varying the Hamiltonian) and to preserve the constraints. Requiring that the time derivatives of this (primary) set of constraints vanish imposes new (secondary) constraints, as expected. But when the secondary constraints are included, one finds that the complete set is preserved. Thus classical supergravity is (in the dynamic sense) a consistent theory. Note that if torsion were dropped from the theory, or if the spinor fields were commuting fields, then the constraints would not be preserved and one would have to add more and more of them.

The constraints of CSG come in two varieties. Certain of them are, in the language of Dirac [19], "second class", and these can be used directly to eliminate variables from the phase space (at the cost of making the symplectic structure a bit strange). Equation (3.5) is of this type (so we can eliminate $\underset{(\phi)}{p}{}^{i}_{A}$) and the constraints related to the vanishing of the momentum conjugate to $\Lambda^{\hat{\alpha}\hat{\beta}}$ are as well. Included there is a constraint which is essentially the spatial part of equation (1.31); it fixes the spatial part of the torsion in terms of the other variables, and permits one to eliminate $\Lambda^{\hat{\alpha}\hat{\beta}}$ and its momentum from the phase space.

The other constraints (which are first class in Dirac's language) include the vanishing of $\underset{(N)}{p}$, $\underset{(\lambda)}{p}$, and $\underset{(\ell)}{p}$, plus the vanishing of a set of quantities which we shall label as $\underset{\text{Diff}}{H_{\hat{\alpha}}}$, $\underset{\text{Lor}}{H_{\hat{\alpha}\hat{\beta}}}$, $\underset{\text{S}}{H_{A}}$ and refer to collectively as the super-Hamiltonians. These super-Hamiltonians are functionals** of $E^{\hat{\alpha}}$, ϕ^{A}, and $\pi_{\hat{\alpha}}$ (the momentum conjugate to $E^{\hat{\alpha}}$). In terms of them, we can write the Hamiltonian for CSG as

$$H = \int_{\Sigma}\Big\{\Big[\ N^{\hat{\alpha}}\ \underset{\text{Diff}}{H_{\hat{\alpha}}} + \ell^{\hat{\alpha}\hat{\beta}}\ \underset{\text{Lor}}{H_{\hat{\alpha}\hat{\beta}}} + \lambda^{A}\ \underset{\text{S}}{H_{A}}\Big] + \underset{(N)}{\mu}{}^{\hat{\alpha}}\ \underset{(N)}{p}{}_{\hat{\alpha}} + \underset{(\ell)}{\mu}{}^{\hat{\alpha}\hat{\beta}} + \underset{(\ell)}{p}{}_{\hat{\alpha}\hat{\beta}}\ \underset{(\lambda)}{\mu}{}^{A}\ \underset{(\lambda)}{p}{}_{A} \tag{3.6}$$

* One can show that n^{μ} is an algebraic functional of $E^{\hat{\alpha}}{}_i$.

** See [24] for explicit formulae.

where $\underset{(N)}{\mu}^{\hat{\alpha}}$, $\underset{(\ell)}{\mu}^{\hat{\alpha}\hat{\beta}}$, and $\underset{(\lambda)}{\mu}^{A}$ are Lagrange multipliers. This form for the Hamiltonian is very important and should be no surprise to anyone familiar with the Hamiltonian formulation of the Einstein or Einstein-Yang-Mills theory (see, e.g. [25] and[20]). It points out the roles of $N^{\hat{\alpha}}$, $\ell^{\hat{\alpha}\hat{\beta}}$ and λ^A as atlas fields ($N^{\hat{\alpha}}$ controlling spacetime diffeomorphisms, $\ell^{\hat{\alpha}\hat{\beta}}$ controlling local Lorentz rotations, and λ^A controlling supertransformations) the roles of the super Hamiltonians as transformation-generating functions (H_{Diff} for spacetime diffeomorphisms, H_{Lor} for Lorentz rotations, and H_{S_A} for supertransformations) and finally it isolates $E^{\hat{\alpha}}$, $\pi_{\hat{\beta}}$, and ϕ^A as the dynamic fields of supergravity. Based on these rôles, one can write the evolution equations for the dynamic fields in "adjoint-form"

$$\partial_t \begin{pmatrix} E \\ \pi \\ \phi \end{pmatrix} = J^{-1} \mathcal{D} \begin{pmatrix} H_{\text{Diff}} \\ H_{\text{Lor}} \\ H_S \end{pmatrix}^{*} \cdot \begin{pmatrix} N \\ \ell \\ \lambda \end{pmatrix} \qquad (3.7)$$

(where J is a sympletic matrix of nontrivial form*, and $\mathcal{D}(\ )^{*}$ is a matrix of differential operators obtained by taking the L^2 adjoint of the Frechét differential of the super-Hamiltonians). This is the first step in carrying out the Fischer-Marsden-Moncrief analysis of the space of solutions (see Bao [24]).

We now show that the Cauchy problem for classical supergravity is formally well-posed. As in Maxwell's theory, it is useful to work with the spacetime covariant formulation of supergravity: So we work with the fields $e^{\hat{\alpha}}_{\ \mu}$ and $\psi^A_{\ \mu}$, and the field equations

$$G^{\mu}_{\ \hat{\alpha}} = \tfrac{1}{2}\varepsilon^{\mu\lambda\nu\rho}\,\bar{\psi}_\lambda\,\gamma_5\,\gamma_{\hat{\alpha}}\,D_\nu\,\psi_\rho \qquad (1.26)$$

and $$0 = i_\gamma\, D\psi = \gamma^{\mu}(D_\mu\psi_\lambda - D_\lambda\psi_\mu) \qquad (1.30)$$

(We shall treat the torsion equation (1.28), and its corollary (1.31) as identities). We keep in mind, however, that $E^{\hat{\alpha}} = e^{\hat{\alpha}}_{\ i}\, dx^i$, $\phi^A = \psi^A_{\ i}\, dx^i$ and $\pi_{\hat{\alpha}}$ (specified at "t = 0") constitute a complete set of initial data; while $N^{\hat{\alpha}} = e^{\hat{\alpha}}_{\ 0}$ and $\ell^{\hat{\alpha}}_{\ \hat{\beta}} = \Gamma^{\hat{\alpha}}_{\ \hat{\beta}0}$ and $\lambda^A = \lambda^A_{\ 0}$ are atlas fields the values and time derivatives of which are independent of the initial data and may be chosen at one's discretion.

---

* The complications in J stem from our using the second class constraints such as (3.5) to eliminate variables in phase space.

The supergravity field equations (1.26) and (1.30) are formally a mixed-order system of partial differential equations. That is, while second derivatives of the tetrad field appear, only first derivatives of the spinor field appear. There are definitions of hyperbolicity and theorems of well-posedness for such systems [21]. However, we find it more convenient to prove well-posedness of CSG by using a purely second order system as an intermediary.

The second order system we work with consists of equation (1.26) together with the equation

$$(i_\gamma D + D i_\gamma)(i_\gamma D + D i_\gamma)\psi = 0 \ . \tag{3.9}$$

The idea is to first show that this second order system (which we shall refer to as CSG II) it itself well-posed, and then to show that spacetime solutions of the CSG equations (1.26) and (1.30) can be generated from supergravity initial data evolved via CSG II.

The strategy we use to prove that the Cauchy problem for CSG II is well-posed is much like that used for Maxwell's theory (see § 2). That is, we look for gauge conditions on the fields $e^{\hat{\alpha}}{}_\mu$ and $\psi^A{}_\mu$ which make CSG II exactly equivalent to a manifestly hyperbolic system, and then we prove that the gauge conditions chosen are compatible with this manifestly hyperbolic system (in the sense that data which initially satisfies the gauge conditions and is evolved by the hyperbolic system will continue to satisfy the gauge condition). In carrying this strategy out, it is important to note that the gauge freedom of CSG II is <u>not</u> the same as that of CSG. While both systems are covariant under spacetime diffeomorphisms and Lorentz frame rotations, only the supergravity system (1.26) and (1.30) is covariant under supersymmetry transformations. We also point out that the Cauchy problem for CSG II is much like that of CSG, but with a few important differences: The atlas fields for CSG II include $N^{\hat{\alpha}} = e^{\hat{\alpha}}{}_0$ and $\ell^{\hat{\alpha}\hat{\beta}} = \Gamma^{\hat{\alpha}\hat{\beta}}{}_0$ but <u>not</u> $\lambda^A = \psi^A{}_0$. A complete set of initial data for CSG II includes $E^{\hat{\alpha}}$, $\pi_{\hat{\beta}}$, ϕ^A, <u>and</u> λ^A, $\partial_t\phi^A$ and $\partial_t\lambda^A$. The constraints for CSG II include modified versions of $H_{\text{Diff}} = 0$ and $H_{\text{Lor}} = 0$, but <u>not</u> $H_S = 0$.

The proper gauge conditions to choose for CSG II are suggested by the following re-expression of its field equations:

$$\tilde{R}_{\mu\nu} = 0_1 \tag{3.10}$$

$$(i_\gamma \tilde{D} + \tilde{D} i_\gamma)(i_\gamma \tilde{D} + \tilde{D} i_\gamma)\psi = 0_1 \tag{3.11}$$

(Here $\tilde{D}$ and $\tilde{R}$ are the covariant derivative and the curvature <u>without torsion</u>, and 0_1 is a generic symbol used to refer to terms involving no higher than first derivatives of any field). What we have done here is notice that the torsion terms contain no derivatives, while the stress-energy of (1.26) (namely $\varepsilon^{\mu\lambda\nu\rho}\,\bar{\psi}_\lambda\,\gamma_5\gamma_{\hat{\alpha}}D_\nu\psi_\rho$) involves only first derivatives; so we have collected such terms into the generic catch-all 0_1. Since, as suggested by the Leray theorem, one need only keep track of terms involving second derivatives to verify hyperbolicity, the equations (3.10) - (3.11) are effectively decoupled. Equation (3.10), in fact, is essentially just the vacuum Einstein equation. It is well-known [26] how to handle that: If one imposes the harmonic gauge condition

$$g^{\mu\nu}\{^{\lambda}_{\ \mu\nu}\} = 0 \tag{3.12}$$

then (3.10) becomes

$$-\frac{1}{2}\,\partial^2\, g_{\mu\nu} = 0_1 \tag{3.13}$$

where "∂^2" is defined by

$$\partial^2 := g^{\mu\nu}\,\partial_\mu\,\partial_\nu = \partial^\mu\,\partial_\mu\;. \tag{3.14}$$

This is the desired hyperbolic operator form on the metric. But in supergravity, the field of interest is the tetrad. Hence (3.13) should be written instead as

$$e_{\hat{\beta}(\mu}\,\partial^2\, e^{\hat{\beta}}{}_{\nu)} = 0_1\;. \tag{3.15}$$

(This follows easily from the identity $g_{\mu\nu} = e_{\hat{\beta}(\mu}e^{\hat{\beta}}{}_{\nu)}$.) To get this into hyperbolic operator form on $e^{\beta}{}_{\nu}$, we need another gauge condition. We impose the requirement

$$\partial^\mu\,\Gamma^{\hat{\alpha}\hat{\beta}}{}_{\mu} = 0 \tag{3.16}$$

which we call the "0(3,1) gauge condition". (Note its superficial similarity to the Lorentz gauge condition $\partial^\mu A_\mu = 0$ which is used to prove that Maxwell's theory is well-posed). The usefulness of the 0(3,1) gauge condition is not evident until one rewrites $\Gamma^{\alpha\beta}{}_{\mu}$ in terms of $e^{\alpha}{}_{\mu}$ and its derivatives (see equation (1.8)) and imposes the harmonic condition (3.12) on the result. Then (3.16) becomes

$$e_{\hat{\beta}[\mu}\partial^2 e^{\hat{\beta}}{}_{\nu]} = 0\;. \tag{3.17}$$

Combining this with (3.15), we get the desired form

$$\partial^2\; e^{\hat{\beta}}{}_{\mu} = 0_1\;. \tag{3.18}$$

It would be very nice if the gauge conditions just chosen would render (3.11) equivalent to the equation $\partial^2\;\psi = 0_1$. But this does not happen. Instead, we find that (3.11) together with gauge conditions (3.12) and (3.16) imply

$$\partial^{\hat{\alpha}} \psi^{A}{}_{\mu} - R^{\lambda}{}_{\mu\nu\xi}\, \sigma^{\nu\xi}{}^{A}{}_{B}\psi^{B}{}_{\lambda} = 0_{1} \tag{3.19}$$

(The calculations which produce (3.19) are quite long and tedious. They are summarized in [24]). This is not in the standard diagonal hyperbolic operator form (2.10) because the Riemann curvature tensor involves second derivatives of the tetrad. Rather, the complete system takes the form

$$\begin{pmatrix} \partial^2 & 0 \\ B^{\mu\nu}\,\partial_\mu\,\partial_\nu & \partial^2 \end{pmatrix}\begin{pmatrix} E \\ \psi \end{pmatrix} = 0_{1} \tag{3.20}$$

(with $B^{\mu\nu}$ being some messy function of the tetrad and spinor fields)

Fortunately, as shown by Choquet-Bruhat [27], the Leray-Ohya [28] theory of nonstrictly hyperbolic partial differential equation systems can handle a triangular system like (3.20). One must verify certain order conditions on the symbol of the system (3.20); but if these are satisfied (see [24] for a verification that they are) then well-posedness follows. So does causal propagation with the causal cone structure determined by the Lorentz metric appearing on the diagonal terms - namely $g_{\mu\nu}$.

We have now shown that for fields $\{e^{\hat{\alpha}}{}_{\mu}, \psi^{A}{}_{\mu}\}$ satisfying the gauge conditions (3.12) and (3.16), the CSG II field equations (1.26) and (3.9) are exactly equivalent to the manifestly well-posed system (3.20). To use this equivalence to prove that CSG II is well-posed, we need to show that the harmonic and O(3,1) gauge conditions are compatible with (3.20).

For the harmonic gauge condition, the verification of compatibility with (3.20) is an old story. This is because (3.20) implies (3.13) which is exactly the hyperbolic equation for the metric which occurs in Einstein's theory with any source field, if the harmonic gauge is imposed. So we simply follow the standard argument (see Choquet-Bruhat [26]): One uses (3.13) plus the Bianchi identity $\nabla^{\mu} G^{\hat{\alpha}}{}_{\mu} = 0$ to derive

$$\partial^2 [g^{\mu\nu}\{^{\lambda}{}_{\mu\nu}\}] = 0_{1}(g^{\mu\nu}\{^{\lambda}{}_{\mu\nu}\}) \tag{3.21}$$

(where $0_1(Z)$, for any quantity Z, denotes a set of terms which vanishes when $Z = 0$ and $\partial_{\mu} Z = 0$, and which involve no higher than first derivatives of Z). One notices that the condition $g^{\mu\nu}\{^{\lambda}{}_{\mu\nu}\}|_{t=0} = 0$ can be solved for $\partial_t e^{\hat{\alpha}}{}_0$ (which is purely atlas field in CSG II as well as in supergravity). And one finds that $g^{\mu\nu}\{\lambda_{\mu\nu}\}|_{t=0} = 0$ together with the CSG II version of the H_{Diff} constaint automatically implies that $\partial_t[g^{\mu\nu}\{\lambda_{\mu\nu}\}]|_{t=0} = 0$. Hence since equation (3.21) has unique solutions and since $g^{\mu\nu}\{\lambda_{\mu\gamma}\} = 0$ is a solution compatible with the initial data, the harmonic condition is compatible with (3.20).

As for the O(3,1) gauge condition, $\partial^\mu \Gamma^{\hat{\alpha}\hat{\beta}}{}_\mu = 0$, its compatibility with (3.20 is easily verified. Indeed, we recall that if the harmonic condition is satisfied, then $\partial^\mu \Gamma^{\hat{\alpha}\hat{\beta}}{}_\mu = 0$ is equivalent to (3.17). But (3.17) has been built directly into the system (3.20). So compatibility is automatic.

This completes the proof that the purely second order system consisting of equations (1.26) and (3.9) is well-posed and has causal propagation. We now return to the field equations of supergravity (1.26) and (1.30). Again, our strategy is to show that a set of initial data for CSG can be evolved via the CSG II evolution equations in such a way that the CSG field equations (2.16) and (1.30) are satisfied throughout the evolved spacetime.

Of course CSG and CSG II share equation (1.26) so we focus our attention on comparing (1.30) and (3.9). And CSG shares diffeomorphism and Lorentz-frame covariance with CSG II, so we focus on finding a gauge for supersymmetry covariance (not shared by CSG II).

Note that if a set of fields $e^{\hat{\alpha}}$ and ψ satisfy the CSG field equation $i_\gamma D\psi = 0$ and if in addition they satisfy the condition

$$0 = i_\gamma \psi = \gamma^\mu \psi_\mu \tag{3.22}$$

then the fields will satisfy the CSG II field equation $(i_\gamma D + D i_\gamma)(i_\gamma D + D i_\gamma)\psi = 0$. This fact is not directly relevant to our argument but it does point out an appropriate gauge condition to fix the supersymmetry gauge freedom: the Rarita-Schwinger gauge condition (3.22).

Let $\{E^{\hat{\alpha}}, \pi_{\hat{\beta}}, \phi\}|_{t=0}$ be a set of initial data for CSG. We want to show that we can extend this to Cauchy data $\{E^{\hat{\alpha}}, \pi_{\hat{\beta}}, \phi, \lambda, \partial_t\lambda\}|_{t=0}$ for CSG II such that

$$i_\gamma \psi|_{t=0} = 0 \tag{3.23}$$

$$\partial_t(i_\gamma \psi)|_{t=0} = 0 \tag{3.24}$$

$$i_\gamma D\psi|_{t=0} = 0 \tag{3.25}$$

$$\partial_t(i_\gamma D\psi)|_{t=0} = 0 \tag{3.26}$$

are all satisfied. Equations (3.23) and (3.24) can be solved for λ and $\partial_t\lambda$, respectively, so they are fine. Equation (3.25) can be solved for $\partial_t\phi$, so it is fine as well. As for (3.26) we note that if one substitutes (3.23) - (3.25) into $(i_\gamma D+Di_\gamma)(i_\gamma D+Di_\gamma)\psi|_{t=0}=0$, then (3.26) is a necessary consequence.

Now let this extended data be evolved via the (well-posed) CSG II system (1.26) and (3.9). Does the CSG equation (3.25) remain satisfied? To show that it does, we define a quantity

$$\Lambda := (i_\gamma D+Di_\gamma)\psi. \tag{3.27}$$

Clearly if the Rarita-Schwinger gauge condition (3.22) holds, then $\Lambda = 0$ iff $i_\gamma D\psi = 0$. Now the CSG II equation (3.9) can be written as

$$(i_\gamma D+Di_\gamma)\Lambda = 0. \tag{3.28}$$

Applying the operator $(i_\gamma D+Di_\gamma)$ to (3.28), we get

$$(i_\gamma D+Di_\gamma)(i_\gamma D+Di_\gamma)\Lambda = 0. \tag{3.29}$$

But since Λ is an anticommuting Majorana spinor-valued one-form (just like ψ), the same calculation which transforms (3.9) into (3.19) turns (3.29) into*

$$\partial^2\,\Lambda = 0_1(\Lambda). \tag{3.30}$$

This equation, together with the initial data $\Lambda|_{t=0} = 0$ (from eqs (3.23)and (3.25)) and $\partial_t\Lambda|_{t=0} = 0$ (from eqs (3.24) and (3.26)) imply that $\Lambda = 0$ throughout the evolved spacetime. But $\Lambda = 0$ implies $i_\gamma D\psi = 0$ if $i_\gamma\,\psi = 0$ holds. And the initial data (3.23) - (3.24) together with the result**

$$0 = D_\lambda(\varepsilon^{\lambda\mu\nu\rho}\gamma_\mu D_\nu\psi_\rho) = -\frac{1}{2}\,\partial^2(i_\gamma\psi) + 0_1(i_\gamma\psi,\Lambda) \tag{3.31}$$

implies that $i_\gamma\psi = 0$ is satisfied throughout spacetime. Hence the evolution by (1.26) and (3.9) does produce a solution of (1.26) and (1.30).

* Yes, there is a curvature term in (3.30) as in (3.19); but it may be subsumed into the catch-all term 0_1.

** In obtaining (3.31), we use the conservation law which states that the divergence of the Rarita-Schwinger equation (1.27) is zero (as long as the Einstein equation (1.26) is satisfied).

This concludes our formal proof that the Cauchy problem for classical supergravity is well-posed and has causal propagation.

Note that in stating and proving our results here, we have said nothing about uniqueness. In Einstein's theory and Maxwell's theory, one can show that for a given set of Cauchy data, the evolved spacetime solution is unique up to diffeomorphism or gauge transformation. But for CSG, we have no such result since little is known about finite supersymmetry transformations. This same lack of knowledge prevents one from showing (as in Einstein's theory or Maxwell's theory) that every globally hyperbolic solution can be generated by evolution from some set of Cauchy data.

There is another, more serious problem with the results we have discussed here: We remind the reader that we have yet to give a complete mathematical definition of the supergravity fields $e^{\hat{\alpha}}_{\mu}$ and $\psi^A{}_{\mu}$ - one of which takes into account the anticommuting nature of $\psi^A{}_{\mu}$. Strictly speaking, the Leray theorems we have cited can only be applied to fields which live in suitable function spaces. So our results are only formal.

In the next section we address this problem of defining the fields more precisely.

§ 4 THE EXTERIOR ALGEBRA FORMULATION OF CLASSICAL SUPERGRAVITY.

The field theory of supergravity is inconsistent unless the spinor fields ψ have anti-commuting components $\psi^A{}_{\mu}$. For quantum field theory, in which the fields are operators on a Hilbert space, anti-commuting fields are a familiar and easily-handled phenomenon. This is not the case with classical field theory, in which the fields are supposed to be tensor fields on spacetime or sections of bundles over spacetime. In this section, we describe a way (See [29], [30], [31], [32], [33], [34]) by which the classical fields can be made to anti-commute, with the CSG field equations becoming a system of partial differential equations on elements of some function space, as desired. But we shall see that this procedure - which is to have the fields take values in an exterior algebra - leads to properties which are suspect on physical grounds.

Before discussing the exterior algebra approach and its implications, we wish to review the properties which it, or any alternative scheme, must account for: The components of the tetrad field $e^{\hat{\alpha}}{}_{\mu}$ and of the spinor field $\psi^A{}_{\mu}$ must have values in an algebra, which we call $\mathfrak{A}$ (so that addition and multiplication make sense as in the formulae $g_{\mu\nu} = \eta_{\hat{\alpha}\hat{\beta}}\, e^{\hat{\alpha}}{}_{\mu}\, e^{\hat{\beta}}{}_{\nu}$ and $Q^{\xi}{}_{\mu\nu} = -\frac{1}{4}\, \bar{\psi}_{\mu} e^{\xi}_{\hat{\alpha}}\gamma^{\hat{\alpha}}\psi_{\nu}$). This algebra must have a unit element (so that invertibility of the tetrad $e^{\hat{\alpha}}{}_{\mu}$ and the metric $g_{\mu\nu}$ make sense), and $\mathfrak{A}$ must be an algebra over the reals (since in the Majorana representation, the

$\gamma^{\hat{\alpha}}$ matrices are real and must multiply the components of $\psi^A{}_\mu$ and $e^{\hat{\alpha}}{}_\mu$). The components of $e^{\hat{\alpha}}{}_\mu$ must commute with each other and should have values in a commutative subalgebra called $\mathfrak{A}^+$. The components of $\psi^A{}_\mu$ must anticommute with each other and so should have values in an anticommutative subvector space called $\mathfrak{A}^-$. However, $\mathfrak{A}^-$ does not form an algebra since the product of a pair of spinor fields (e.g., $\bar{\psi}_\mu \gamma^{\hat{\alpha}} \psi_\nu$) no longer anti-commutes with other spinor fields. In fact, such quadratic functions of spinor fields commute, so they may be viewed as being elements of $\mathfrak{A}^+$. We also note that the spinor fields commute with the tetrad fields.

The properties thus far described are exactly those of a graded-commutative $\mathbb{Z}_2$-graded algebra* $\mathfrak{A}$ over the reals, with a unit element. The tetrad field components $e^{\hat{\alpha}}{}_\mu$ take values in $\mathfrak{A}^+$ while the spinor field components take values in $\mathfrak{A}^-$.

We need to include two more properties. Firstly, to permit an arbitrarily large number of linearly independent local supersymmetry transformations (of the type given by equation (1.32)), we require that $\mathfrak{A}$ be infinite dimensional. Secondly, so that bona fide pde theorems (of the Leray type) may be applied to CSG, we require that $\mathfrak{A}$ admits a Banach norm (with the multiplication operation on $\mathfrak{A}$ being continuous with respect to this norm).

One way to build an algebra with all of the properties just described is as follows (See references [29]-[34]). We start with a real, infinite dimensional Banach space W. On W, we define an exterior ("wedge") product $\wedge$ (continuous with respect to the Banach norm) with the usual properties (e.g., $\alpha \wedge \beta = -\beta \wedge \alpha$ for $\alpha, \beta \in W$). Then we define

$$\begin{aligned}
\mathfrak{A}_0 &:= \mathbb{R} \qquad (4.1)\\
\mathfrak{A}_1 &:= W\\
\mathfrak{A}_2 &:= W \wedge W := \{\sum_{ij} \alpha_i \wedge \beta_j \,|\, \alpha_i, \beta_j \in W\}\\
\mathfrak{A}_3 &:= W \wedge W \wedge W
\end{aligned}$$

etc.

The algebra $\mathfrak{A}$ is just the direct sum of all the $\mathfrak{A}_n$-spaces; i.e.

$$\mathfrak{A} := \bigoplus_{n=0}^{\infty} \mathfrak{A}_n \,. \qquad (4.2)$$

* A $\mathbb{Z}_2$-graded algebra is defined to be the direct sum of an even part $\mathfrak{A}^+$ and an odd part $\mathfrak{A}^-$ such that $\mathfrak{A}^+\mathfrak{A}^+ \subset \mathfrak{A}^+$, $\mathfrak{A}^-\mathfrak{A}^- \subset \mathfrak{A}^+$, $\mathfrak{A}^+\mathfrak{A}^- \subset \mathfrak{A}^-$ and $\mathfrak{A}^-\mathfrak{A}^+ \subset \mathfrak{A}^-$. It is graded commutative if two even elements commute, two odd elements anti-commute, and an even element commutes with an odd element.

The even and odd parts are then the direct sums for even and odd n, respectively:

$$\mathcal{A}^{+} := \bigoplus_{n=0}^{\infty} \mathcal{A}_{2n} \tag{4.3a}$$

$$\mathcal{A}^{-} := \bigoplus_{n=0}^{\infty} \mathcal{A}_{2n+1} \tag{4.3b}$$

Thus $\mathcal{A} = \mathcal{A}^{+} \oplus \mathcal{A}^{-}$, and all of the properties mentioned above are easily verified.

This construction is known as the exterior algebra over W; notationally one writes $\mathcal{A} = \Lambda^{*} W$. Note that $\Lambda^{*} W$, in addition to being $\mathbb{Z}_2$-graded (even part $\mathcal{A}^{+}$ and odd part $\mathcal{A}^{-}$) is $\mathbb{Z}^{+}$-graded (integer parts $\mathcal{A}_n$). Note also that if $\{\alpha^{\Phi}\}$ forms a basis for W, then $\{\alpha^{\Phi}\wedge\alpha^{\Psi}\}$ forms a basis for $\mathcal{A}_2$, $\{\alpha^{\Phi}\wedge\alpha^{\Psi}\wedge\alpha^{\Theta}\}$ forms a basis for $\mathcal{A}_3$, etc.

We now describe what the formulation of classical supergravity based on $\mathcal{A} = \Lambda^{*} W$ (we shall abbreviate it as CSG $\Lambda^{*} W$) looks like. The tetrad field $e^{\hat{\alpha}}$ is an $(\mathbb{R}^4 \otimes \mathcal{A}^{+})$-valued one-form which can be written as

$$\begin{aligned} e^{\hat{\alpha}} &= (e^{\hat{\alpha}}{}_{\mu} + e^{\hat{\alpha}}{}_{\Phi\Psi\mu}\, \alpha^{\Phi}\wedge\alpha^{\Psi} + e^{\hat{\alpha}}{}_{\Phi\Psi\Theta\Sigma\mu}\, \alpha^{\Phi}\wedge\alpha^{\Psi}\wedge\alpha^{\Theta}\wedge\alpha^{\Sigma} + \ldots)dx^{\mu} \\ &= (\underset{0}{e}{}^{\hat{\alpha}}{}_{\mu} + \underset{2}{e}{}^{\hat{\alpha}}{}_{\mu}\, \alpha\wedge\alpha + \underset{4}{e}{}^{\hat{\alpha}}{}_{\mu}\, \alpha\wedge\alpha\wedge\alpha\wedge\alpha + \ldots)dx^{\mu} \end{aligned} \tag{4.4}$$

The spinor field ψ^{A} is a (Majorana spinor $\otimes\, \mathcal{A}^{-}$)-valued one-form which can be expressed as

$$\psi^{A} = (\underset{1}{\psi}{}^{A}{}_{\mu}\, \alpha + \underset{3}{\psi}{}^{A}{}_{\mu}\, \alpha\wedge\alpha\wedge\alpha + \ldots)dx^{\mu}. \tag{4.5}$$

Note that the components here - e.g. $\underset{0}{e}{}^{\hat{\alpha}}{}_{\mu}$, $\underset{1}{\psi}{}^{A}{}_{\mu}$, etc. - are all simple (commuting) functions.

If we now plug these fields into the supergravity field equations (1.26) and (1.30), they too take values in $\mathcal{A}$. Schematically, we obtain

$$\begin{aligned} 0 &= G - \bar{\psi}\mathcal{D}\psi \\ &= \{\underset{0}{G}(\underset{0}{e})\} + \{\underset{2}{G}(\underset{0}{e},\underset{1}{\psi},\underset{2}{e}) - \underset{1}{\bar{\psi}}\,\underset{0}{\mathcal{D}}(\underset{0}{e})\underset{1}{\psi}\}\alpha\wedge\alpha \\ &\qquad + \{\underset{4}{G}(\underset{0}{e},\underset{1}{\psi},\underset{2}{e},\underset{3}{\psi},\underset{4}{e}) - \underset{1}{\bar{\psi}}\,\underset{0}{\mathcal{D}}(\underset{0}{e})\underset{3}{\psi} - \underset{3}{\bar{\psi}}\,\underset{0}{\mathcal{D}}(\underset{0}{e})\underset{1}{\psi} - \underset{1}{\bar{\psi}}\,\underset{2}{\mathcal{D}}(\underset{0}{e},\underset{1}{\psi},\underset{2}{e})\underset{1}{\psi}\}\alpha\wedge\alpha\wedge\alpha\wedge\alpha \\ &\qquad + \ldots \end{aligned} \tag{4.6}$$

and

$$0 = \mathfrak{D}\psi \tag{4.7}$$

$$= \{\underset{0}{\mathfrak{D}}(\underset{0}{e})\underset{1}{\psi}\}\alpha + \{\underset{0}{\mathfrak{D}}(\underset{0}{e})\underset{3}{\psi} + \underset{2}{\mathfrak{D}}(\underset{0}{e},\underset{1}{\psi},\underset{2}{e})\underset{1}{\psi}\}\alpha\wedge\alpha\wedge\alpha + \ldots$$

where $\underset{n}{G}(\underset{0}{e},\underset{1}{\psi},\underset{2}{e},\underset{3}{\psi},\ldots\underset{n}{e})$ denotes the component of the piece of the Einstein tensor which is contained in $\mathfrak{A}_n$, and is constructed out of $\underset{0}{e},\underset{1}{\psi},\underset{2}{e},\underset{3}{\psi},\ldots\underset{n}{e}$.* (Note that G_n (for $n \neq 0$) is linear in $\underset{n}{e}$) and $\underset{n}{\mathfrak{D}}(\underset{0}{e},\underset{1}{\psi},\underset{2}{e},\ldots\underset{n}{e})$ denotes some appropriate $\gamma \wedge D$ combination which is contained in $\mathfrak{A}_n$ and is constructed out of $\underset{0}{e},\underset{1}{\psi},\underset{2}{e},\underset{3}{\psi},\ldots\underset{n}{e}$ (also linear in $\underset{n}{e}$). Now if these equations are to be satisfied, then <u>order-by-order</u> the right-hand-sides of (4.6) and (4.7) must vanish. That is, we must have

$$0 = \underset{0}{G}(\underset{0}{e}) \ , \tag{4.8a}$$

$$0 = \underset{2}{G}(\underset{0}{e},\underset{1}{\psi},\underset{2}{e}) - \underset{1}{\bar{\psi}}\,\underset{0}{\mathfrak{D}}(\underset{0}{e})\underset{1}{\psi}, \tag{4.8b}$$

etc.

and

$$0 = \underset{0}{D}(\underset{0}{e})\underset{1}{\psi} \ , \tag{4.9a}$$

$$0 = \underset{0}{D}(\underset{0}{e})\underset{3}{\psi} + \underset{2}{D}(\underset{0}{e},\underset{2}{e})\underset{1}{\psi} \ , \tag{4.9b}$$

etc.

In equations (4.8) - (4.9), we have the field equations for CSG - Λ* W, expressed as partial differential equations on every-day functions $\underset{0}{e}$, $\underset{1}{\psi}$, etc. There are two important features of this infinite sequence of equations: Firstly, they are sequentially decoupled. That is, (4.8a) can be solved for $\underset{0}{e}$, then (4.9a) can be solved for $\underset{1}{\psi}$ (with $\underset{0}{e}$ known), then (4.8b) can be solved for $\underset{2}{e}$, etc. Secondly, all but (4.8a) are <u>linear</u> in the variable to be solved for (i.e., (4.9a) is linear in $\underset{1}{\psi}$, (4.8b) is linear in $\underset{2}{e}$, etc.), while (4.8a) is just the vacuum Einstein equation.

Mathematically, these features are very nice. There is only one nonlinear equation to handle, and this is a familiar one. Such questions as linearization stability and the nature of the space of solutions for this exterior algebra formulation of classical supergravity are settled by looking at the vacuum Einstein equation, for which the answers are known [22]. Well-posedness becomes very simple to prove because the vacuum Einstein equation is known to have a well-posed Cauchy problem, and the rest of

---

* The dependence of G and $\mathfrak{D}$ upon $\underset{1}{\psi},\underset{3}{\psi}$, etc. is the result of the presence of torsion (recall eq. 1.31).

the equations in (4.8) and (4.9) can easily be shown to be hyperbolic, and therefore well-posed, using standard techniques for linear equations [35]. Indeed, in an appropriate gauge, all of the equations except the first in (4.8) become

$$\partial_0^2 e_n = F(e_{n-2}, \ldots, e_0, \psi_{n-1}, \ldots, \psi_1) \qquad (4.10)$$

while those in (4.9) become

$$\mathcal{D}_0 \psi_n = G(\psi_{n-2}, \ldots, \psi_1, e_{n-1}, \ldots, e_0) \qquad (4.11)$$

where $\partial_0^2 := g_0^{\mu\nu} \partial_\mu \partial_\nu$, and $\mathcal{D}_0$ is a first order hyperbolic operator involving only e_0. Note that the characteristics of (4.10) and (4.11) - and therefore the causal cone structure of the theory - are determined completely by e_0.

While the mathematics of the exterior algebra formulation is quite nice, the physics is suspicious. One of the big problems relates to the observation just made: that the causal cone structure of the theory is determined by e_0 alone. Two of the fundamental tenets of the present-day view of gravity are (1) that the causal cone structure of spacetime is determined by the gravitational field, and (2) that the gravitational field is affected by the presence of source fields. Yet in CSG-Λ^* W, the spin-3/2 field is completely irrelevant to the determination of the causal cone structure of a given spacetime. Another suspicious property is the essential linearity of the theory. General relativity* leads us to expect the gravitational field to behave in a very nonlinear way, especially in the presence of source fields. Yet in CSG-Λ^* W, such linearity is present only in the vacuum Einstein equation.

Before discussing (in the next section) the various ways in which one might try to make sense of classical supergravity, in spite of these problems, we wish to note that the exact same sorts of problems arise in other theories which are built to include anti-commuting spinor fields. The Maxwell-Dirac and the Einstein-Dirac field theories can be set up with an anti-commuting spin-1/2 Dirac field; if one attempts to mathematically formulate these theories using an exterior algebra to encode the anti-commutation, then again decoupling and linearization occur (In the Maxwell-Dirac case, the theory becomes completely linear). There is a big difference between these two theories and classical supergravity, however: In the former theories, anti-commutation of the spinor fields is optional. In CSG, it is indispensable.

* which is presently the most experimentally-supported and believed theory of the gravitational field.

§ 5 DOES CLASSICAL SUPERGRAVITY MAKE SENSE?

The exterior algebra formulation of classical supergravity is well-defined mathematically (and has a well-defined Cauchy problem), but it has properties which seem to violate one's physical intuition. How should one react to this problem? We briefly discuss here a few possibilities.

A) Search for Other Formulations:

While $\mathfrak{A} = \Lambda^* W$ has all the properties which we described as necessary for an algebraic formulation of classical supergravity, it may not be the only algebra which does. So, one might seek another choice for $\mathfrak{A}$. To avoid decoupling, it is important that this algebra not be $\mathbb{Z}^+$-graded; $\mathbb{Z}_2$-grading is all one wants. We have not yet found this alternative algebra. Nor have such tricks as defining $\bar{\psi}$ to live in the dual of $\mathfrak{A}^-$ been of any use.

B) Understand the Classical-Quantum Relationship Better:

The properties of CSG-$\Lambda^* W$ which cause physical suspicion arise when one attempts to treat the theory as an everyday classical field theory, involving a source field coupled to Einstein's gravitational field. It may be however, that unlike the Einstein-Maxwell theory, for example, classical supergravity (and any other theory with fermion fields) has no directly classical manifestations. One must then attempt to understand what role, if any, the classical field equations might play in mirroring the quantum behavior of the theory. It is hard to believe that the classical field equations are irrelevant, but their exact role (and the nature of classical limits of quantum field theories) is far from understood.

C) Forget Supergravity:

Nobody has yet seen any sleptons,squarks, or gravitinos. Thus it may be that supergravity and supersymmetry have nothing whatever to do with the physical universe. In that case, the problems discussed here are mathematical oddities and nothing more.

It is not clear to us which, if any, of these possibilities makes the most sense. We are, however, currently pursuing the first two. In favor of these, we remind the reader that formally at least (as shown in § 3) classical supergravity does seem to make sense. Thus we believe its study is worth pursuing.

APPENDIX: NOTATIONS AND CONVENTIONS.

In this appendix, we remark on some of the notations and conventions which are not spelled out in the text of the paper.

We use five types of indices:

(1) $\mu, \nu, \ldots$ (lower case Greek, undecorated). These are spacetime tangent space indices. They range from 0 to 3. They are raised and lowered using the metric $g_{\mu\nu}$.

(2) $\hat{\alpha}, \hat{\beta}, \ldots$ (lower case Greek, with caret). These are the indices labeling the tetrad frame vectors and forms. They range from 0 to 3. They are raised and lowered using the Minkowski metric $\eta_{\hat{\alpha}\hat{\beta}}$.

(3) $i, j, \ldots$ (lower case Latin). These are spatial tangent space indices. They range from 1 to 3. They are raised and lowered using ${}^3g_{ij}$.

(4) $A, B, \ldots$ (upper case Latin). These are the spinor indices. They range from 1 to 4.

(5) $\Phi, \Psi, \ldots$ (upper case Greek) These label the basis elements of the infinite dimensional vector space W, out of which $\mathfrak{a} = \Lambda^* W$ is built. They range from 1 to ∞.

We use five types of derivatives:

(1) ∂_μ. This is the partial derivative (with respect to some coordinate system). It is independent of any connection.

(2) d. This is the exterior differential operator on forms. It is independent of the connection.

(3) ∇_μ. This is the covariant derivative (along ∂_μ), based upon the connection $\Gamma^{\alpha\beta}{}_{\mu}$.

(4) D_μ. This is the derivative operator which corrects (via the connection) for spinor and frame indices, but ignores tangent space indices.

(5) D. This is the covariant exterior differential operator, defined on "___"-valued forms.

We use the operator "i_V", acting on forms, to indicate insertion of the vector V into the first slot of the form it is acting upon. So, e.g., we have $i_V A = V^\mu A_\mu$ for a one form A.

Our conventions on metric signature, formulas for curvatures, etc., follow those of MTW [36].

The notation $T_{(\mu\nu)}$ or $e^{\hat{\alpha}}{}_{(\mu} De_{\hat{\alpha}\nu)}$ indicates symmetrization. That is

$$T_{(\mu\nu)} := \frac{1}{2}\,(T_{\mu\nu}+T_{\nu\mu}) \qquad \text{(A.1)}$$

and

$$e^{\hat{\alpha}}{}_{(\mu}De_{\hat{\alpha}\nu)} := \frac{1}{2}\,(e^{\hat{\alpha}}{}_{\mu}De_{\hat{\alpha}\nu}+e^{\hat{\alpha}}{}_{\nu}De_{\hat{\alpha}\mu}) \; .$$

Similarly

$$T_{[\mu\nu]} := \frac{1}{2}\,(T_{\mu\gamma}-T_{\nu\mu}) . \qquad \text{(A.2)}$$

The gamma matrices $\gamma^{\hat{\alpha}}$ we use are those which generate the Clifford algebra according to

$$\gamma^{\hat{\alpha}}\gamma^{\hat{\beta}} + \gamma^{\hat{\beta}}\gamma^{\hat{\alpha}} = 2\eta^{\hat{\alpha}\hat{\beta}} . \qquad \text{(A.3)}$$

Note that $\gamma^{\hat{\alpha}}$ are spacetime constant (so $\partial_\mu\gamma^{\hat{\alpha}} = 0$). The matrices $\gamma^\mu := \gamma^{\hat{\alpha}}e_{\hat{\alpha}}{}^{\mu}$ are <u>not</u> spacetime constant (so $\partial_\mu\gamma^\nu = \gamma^{\hat{\alpha}}\,\partial_\mu e_{\hat{\alpha}}{}^{\nu}$).

From $\gamma^{\hat{\alpha}}$, one constructs

$$\gamma_5 = \gamma^{\hat{0}}\gamma^{\hat{1}}\gamma^{\hat{2}}\gamma^{\hat{3}} \qquad \text{(A.4)}$$

and $\sigma^{\hat{\alpha}\hat{\beta}} = \frac{1}{4}(\gamma^{\hat{\alpha}}\gamma^{\hat{\beta}} - \gamma^{\hat{\beta}}\gamma^{\hat{\alpha}})$.

We also have $\gamma = \gamma_{\hat{\alpha}}e^{\hat{\alpha}} = \gamma^{\hat{\alpha}}e_{\hat{\alpha}}$ which serves as a vector or a one-form, depending upon the context.

We use the real Majorana representation of the γ matrices, so

$$\gamma^{\hat{0}} = \begin{pmatrix} 0 & 0 & 0 & 1 \\ 0 & 0 & 1 & 0 \\ 0 & -1 & 0 & 0 \\ -1 & 0 & 0 & 0 \end{pmatrix}, \quad \gamma^{\hat{1}} = \begin{pmatrix} 1 & 0 & 0 & 0 \\ 0 & 1 & 0 & 0 \\ 0 & 0 & -1 & 0 \\ 0 & 0 & 0 & -1 \end{pmatrix}, \qquad \text{(A.5)}$$

$$\gamma^{\hat{2}} = \begin{pmatrix} 0 & 0 & 0 & -1 \\ 0 & 0 & 1 & 0 \\ 0 & 1 & 0 & 0 \\ -1 & 0 & 0 & 0 \end{pmatrix}, \quad \gamma^{\hat{3}} = \begin{pmatrix} 0 & 0 & 1 & 0 \\ 0 & 0 & 0 & 1 \\ 1 & 0 & 0 & 0 \\ 0 & 1 & 0 & 0 \end{pmatrix} .$$

In this representation, a Majorana spinor field has real components. Its conjugate $\bar{\psi}$ is then given by $\bar{\psi} = \psi^T\gamma^{\hat{\alpha}}$.

There are a number of identities which are very useful in working with the spinors in supergravity (including the Fierz-rearrangement matrix). Many of these are given in [16].

Acknowledgements: We wish to thank Y. Choquet-Bruhat, J. Nester, I.M. Singer and P. van Nieuwenhuizen for useful discussions. We also thank H.-J. Seifert for his hospitality during the 1983 Conference on supergravity.

REFERENCES

[1] P. van Nieuwenhuizen, D. Freedman, and S. Ferrara: Progress toward a Theory of Supergravity, Phys. Rev. D13 (1976) 3214.

[2] S. Deser and B. Zumino: Consistent Supergravity, Phys. Lett. 62 B (1976) 335.

[3] P. van Nieuwenhuizen: Supergravity, Phys. Rep. 68 (1981), 190.

[4] J. Bagger and J. Wess: *Supersymmetry and Supergravity*, Princeton University Press, (1983).

[5] J. Sherk, in Proceedings of the Stony Brook Conference, North Holland Press (1979).

[6] J.H. Schwarz and J. Sherk: Spontaneous Breaking of Supersymmetry through Dimensional Reduction, Phys. Lett. 82 B (1979) 60.

[7] P. Fayet and J. Iliopoulos: Phys. Lett. 51 B (1974) 461.

[8] M. Grisaru and P. van Nieuwenhuizen: Renormalizability of Supergravity, in *New Pathways in Theoretical Physics* (eds. B.B. Kursonoglu and A. Perlmutter), Coral Gables (1977).

[9] R. Jackiw: Quantum Meaning of Classical Field Theory, Rev. Mod. Phys. 49 (1977), 681.

[10] P.W. Higgs: Spontaneous Symmetry Breaking without Massless Bosons, Phys. Rev. 145 (1966) 1156.

[11] S. Paneitz and I. Segal: Proc. Natl. Acad. Sci. USA, 77, (1980), 6943.

[12] J. Isenberg and J. Nester, Ann. Phys. (NY) 107 (1977) 56.

[13] B. De Witt and P. van Nieuwenhuizen, preprint of book to be published.

[14] D. Bao, J. Isenberg, and P. Yasskin: The Dynamics of the Einstein-Dirac System I: A Principal Bundle Formulation of the Theory and Its Canonical Analysis, preprint.

[15] W. Rarita and J. Schwinger: Phys. Rev. 60 (1941) 61.

[16] D. Bao: Some Aspects in the Dynamics of Supergravity, Ph.D. thesis, Univ. of Cal. (Berkeley), (1983).

[17] S. Deser: From Gravity to Supergravity, in *General Relativity and Gravitation* (ed.: A. Held) Plenum (1980)

[18] J. Isenberg: The Construction from Initial Data of Spacetimes with Nontrivial Spatial and Bundle Topology, Ann. Phys. (N.Y.) 129, (1980) 223.

[19] P. Dirac: *Lectures on Quantum Mechanics*, Academic Press (1964).

[20] J. Isenberg and J. Nester: Canonical Gravity, in *General Relativity and Gravitation* (ed.: A. Held), Plenum (1980).

[21] J. Leray: Hyperbolic Differential Equations, lecture notes, Princeton (1951).

[22] A. Fischer, J. Marsden, and V. Moncrief: The Structure of the Space of Solutions of Einstein's Equations I., Ann. Inst. H. Poincaré 33 (1980) 147.

[23] M. Pilati: The Canonical Formulation of Supergravity and the Quantization of the Ultralocal Theory of Gravity, Ph. D. thesis, Princeton Univ. (1980).

[24] D. Bao: A Sufficient Condition for the Linearization Stability of N = 1 Supergravity: A Preliminary Report, preprint.

[25] K. Kuchař: J. Math. Phys. 17 (1976) 777, 792, 801.

[26] Y. Choquet-Bruhat: The Cauchy Problem, in *Gravitation*, An Introduction to Current Research (ed.:L. Witten), Wiley (1962).

[27] Y. Choquet-Bruhat: Diagonalisation des Systèmes Quasi-Linéaires et Hyperbolicité Non Strictes J. Math. Pures et Appl. 45 (1966), 371.

[28] J. Leray and Y. Ohya: Equations et Systèmes Non Linéaires Hyperboliques Non Strictes,Math. Annalen (1966).

[29] F.A. Berezin and D.A. Leites: Supermanifolds, Sov. Math. Dokl. 16 (1975), 1218.

[30] L. Corwin, Y. Ne'eman, and S. Sternberg: Graded Lie Algebras in Mathematics and Physics (Bose-Fermi symmetry), Rev. Mod. Phys., 47 (1975) 573.

[31] B. Kostant: Graded Manifolds, Graded Lie Theory, and Prequantization, in *Lecture Notes in Math. # 570*, Springer (1977).

[32] M. Batchelor: Supermanifolds, in *Group Theoretical Methods in Physics* (ed.: W. Beiglböck, A. Behm, and E. Takasugi), Springer (1979).

[33] J. Dell and L. Smolin: Graded Manifold Theory as the Geometry of Supersymmetry, Comm. Math. Phys. 66 (1979), 197.

[34] D.A. Leites: Introduction to the Theory of Supermanifolds, Russian Math. Surveys 35:1 (1980) 1.

[35] Y. Choquet-Bruhat: The Cauchy Problem in Classical Supergravity, preprint (1983).

[36] C. Misner, K. Thorne, and J. Wheeler: *Gravitation*, Freeman, (1973).

List of Participants

Dr. N. Backhouse
University of Liverpool, Dept. of Applied Mathematics and Theoretical Physics, P.O. Box 147, Liverpool L69 3BX, U.K.

Dr. M. Batchelor
Dept. of Pure Mathematics and Mathematical Statistics, University of Cambridge, 16 Mill Lane, Cambridge CB2 1SB, U.K.

H. Beyer
Universität zu Köln, Institut für Theoretische Physik, Zülpicher Str. 77, D-5000 Köln 41, FRG.

Dr. P. Bongaarts
Instituut voor theoretische Natuurkunde, Nieuwsteeg 18, 2311 SB Leiden, The Netherlands.

H.-J. Bornkast
Hochschule der Bundeswehr Hamburg, Fachbereich Maschinenbau Holstenhofweg 85, D-2000 Hamburg 70, FRG

Dr. C.J.S. Clarke
University of York, Dept. of Mathematics, Heslinton, York, YO1 5DD, U.K.

Prof. Dr. K.D. Elworthy
University of Warwick, Coventry, Mathematics Institute, Coventry CV 47 AL, U.K.

Dr. H. Friedrich
Hochschule der Bundeswehr Hamburg, Fachbereich Maschinenbau, Holstenhofweg 85, D-2000 Hamburg 70, FRG.

Dr. A. Hirshfeld
Universität Dortmund, Institut für Physik, Postfach 50 05 00, D-4600 Dortmund 50, FRG

Dr. J. Isenberg
University of Oregon, Eugene, OR 97403, U.S.A.
Current address: Dept. of Mathematics, Rice University, P.O. Box 1892, Houston, TX 77251, U.S.A.

Prof. Dr. Müller zum Hagen
Hochschule der Bundeswehr Hamburg, Fachbereich Maschinenbau, Holstenhofweg 85, D-2000 Hamburg 70, FRG.

Dr. J. Nitsch
Universität zu Köln, Institut für Theoretische Physik,
Zülpicher Str. 77, D-5000 Köln 41, FRG.

Dr. J. Rawnsley
Mathematical Institute, University of Warwick,
Coventry CV 4 7AL, U.K.

Dr. A. Rogers
Imperial College of Science and Technology, Blackett Laboratory,
Theoretical Physics Group, London SW7 2B7, U.K.

Dr. A. Rosenblum
Dept. of Physics, Temple University, Philadelphia, Pa 19122, U.S.A.

Dr. R. Salchow
Hochschule der Bundeswehr Hamburg, Fachbereich Maschinenbau,
Holstenhofweg 85, D-2000 Hamburg 70, FRG.

Prof. Dr. H.-J. Seifert
Hochschule der Bundeswehr Hamburg, Fachbereich Maschinenbau,
Holstenhofweg 85, D-2000 Hamburg 70, FRG.

R. Schimmrigk
Universität Heidelberg, Institut für Angewandte Mathematik,
Im Neuenheimer Feld 294, D-6900 Heidelberg 1, FRG.

Dr. B. Schmidt
Max Planck-Institut für Astrophysik, Karl-Schwarzschild-Str. 1
D-8046 Garching bei München, FRG.

Prof. Dr. J. Wess
Universität Karlsruhe, Institut für Theoretische Physik,
Kaiserstraße, D-7500 Karlsruhe 1, FRG.

INDEX

GPSR Compliance

The European Union's (EU) General Product Safety Regulation (GPSR) is a set of rules that requires consumer products to be safe and our obligations to ensure this.

If you have any concerns about our products, you can contact us on ProductSafety@springernature.com

In case Publisher is established outside the EU, the EU authorized representative is:

Springer Nature Customer Service Center GmbH
Europaplatz 3
69115 Heidelberg, Germany

Batch number: 10607222

Printed by Printforce, the Netherlands